Cultural Heritage Science

The preservation and interpretation of our cultural heritage is one of the major challenges of today's society. Cultural Heritage Science is a highly interdisciplinary book series covering all aspects of conservation, analysis and interpretation of artworks, objects and materials from our collective cultural heritage. The series focuses on science and conservation in three main fields

- Art technology
- Active conservation and restoration
- Preventive conservation and risk management

The series addresses conservators and conservation scientists at museums, institutes, universities and heritage organizations. It also provides valuable information for curators and decision makers at museums and heritage organizations. All titles in the book series will be peer reviewed. Titles will be published as printed books and as eBooks, opening up the opportunity to include electronic supplementary material (videos, high-resolution figures, special data formats, and access to databases).

Howell G. M. Edwards • Christopher J. Brooke

Coade Stone

A History and Analysis

Springer

Howell G. M. Edwards
Department of Chemical
and Biosciences
University of Bradford
Bradford, West Yorkshire, UK

Christopher J. Brooke
Department of History
University of Nottingham
Nottingham, UK

ISSN 2366-6226 ISSN 2366-6234 (electronic)
Cultural Heritage Science
ISBN 978-3-031-71431-3 ISBN 978-3-031-71432-0 (eBook)
https://doi.org/10.1007/978-3-031-71432-0

This Springer imprint is published by the registered company Springer Nature Switzerland AG
The registered company address is: Gewerbestrasse 11, 6330 Cham, Switzerland

Howell Edwards dedicates this book to his late wife, Gill, a staunch supporter of his research activities and academic writing over many decades and whose presence for the completion of this work was greatly missed. He acknowledges the many fruitful discussions and support of their daughter, Kate, a geologist and geotechnical engineer, in the telling of a story that is ongoing and has yet many aspects to fulfil.

Acknowledgements

The authors particularly acknowledge the assistance of the following in the research they have carried out to produce this book: Caroline Stanford, The Landmark Trust: Guy Horton-Fawkes, Farnley Hall; Rhona Finlayson, West Yorkshire Archaeology Advisory Services Heritage Environmental Research Officer; Dr Alex Surtees, University of Bradford; Maggie Fleming; Dr Marcello Di Bonito, Nottingham Trent University; Dr David Carrington, Skillington Workshop; Ian George, Lincolnshire County Council; Keith Holland, Lincoln Castle, Tracey Williams, Library Specialist, Heritage and Local Studies, The Core, Solihull, Ian Gow, Architectural Historian, Edinburgh.

Contents

About the Authors

Howell G. M. Edwards M.A., B.Sc., D. Phil., C. Chem., FRSC is Professor Emeritus of Molecular Spectroscopy at the University of Bradford. He read Chemistry at Jesus College in the University of Oxford, and after completing his B.A. and B.Sc. degrees, he studied for his doctorate in Raman spectroscopy at Oxford and then became a Research Fellow at Jesus College, University of Cambridge. He joined the University of Bradford as a Lecturer in Structural and Inorganic Chemistry, becoming Head of the Department of Chemical and Forensic Sciences, and was awarded a Personal Chair in Molecular Spectroscopy in 1996. He has received several international awards (Sir Harold Thompson Award; Charles Mann Award; Emanuel Boricky Medal; Norman Sheppard Award) in a spectroscopic career which has resulted in the publication of 1400 research papers in Raman spectroscopy and the characterisation of materials, along with six books on the application of this analytical technique to art, archaeology and forensic science. He has had a lifelong interest in the porcelains of William Billingsley, especially those from the Derby, Nantgarw and Swansea factories and has authored eight books on porcelain: *Swansea and Nantgarw Porcelains: A Scientific Reappraisal*; *Nantgarw and Swansea Porcelains: An Analytical Perspective*; *Porcelain to Silica Bricks: The Extreme Ceramics of William Weston Young, 1776–1847*; *18th and 19th Century Porcelain Analysis: A Forensic Provenancing Assessment*; *Porcelain Analysis and Its Role in the Forensic Attribution of Ceramic Specimens*; *Raman Spectroscopy in Cultural Heritage Preservation* (with Philippe Colomban and Peter Vandenabeele, Springer, 2023); and *Welsh Armorial Porcelains: Nantgarw and Swansea Crested China*, published by Springer Publishing, Dordrecht, The Netherlands, and two specialist texts, namely *The Pendock-Barry Porcelain Service: A Forensic Evaluation* (with R Denyer and M Denyer, 2023) and *The Farnley Hall Service: A Unique Survivor in Nantgarw Porcelain* (2023). He has also produced several monographs: *William Billingsley – The Enigmatic Porcelain Artist, Decorator and Manufacturer*;

Nantgarw Porcelain: The Pursuit of Perfection (with M.D. Denyer); *Swansea Porcelain – The Duck-Egg Translucent Vision of Lewis Dillwyn*; and *Derby Porcelain: The Golden Years, 1780–1830.* Howell Edwards is Honorary Scientific Adviser to the de Brecy Trust on the scientific evaluation of their artworks and paintings.

Christopher J. Brooke B.Sc., Ph.D., FSA, FSAScot, FRHistS, AFRSPSoc, MCIfA, IHBC is Honorary Associate Professor of Medieval History and Church Archaeology at the University of Nottingham and Honorary Visiting Research Fellow in Remote Sensing at Nottingham Trent University. He read Archaeological Sciences at the University of Bradford, and following completion of his B.Sc. degree he researched for his doctorate in Archaeological Remote Sensing at the University of Nottingham. His career spans periods in higher education, private practice, local and central government, and several years in senior management with the National Trust for Scotland. He has throughout his career maintained various honorary positions at the Universities of Nottingham and Durham and more recently has joined Nottingham Trent University. He has published over 50 research papers and 2 books, mainly in the fields of archaeological remote sensing, spectroscopy, and medieval history and archaeology. His book on defensible churches on the Anglo-Scottish Border: *Safe Sanctuaries, Security and Defence in Anglo-Scottish Border Churches 1290–1690* (John Donald, 2000) remains the definitive work on the subject. He serves in an advisory capacity on several professional bodies.

Chapter 1
Introduction—Coade Artificial Stone and Its Marks

Abstract The origins of the high-fired ceramic Coade stone from Eleanor Coade's taking over Daniel Pincot's ailing manufactory in 1769, through William Croggon's proprietorship after her death in 1821 to its eventual closure in 1838 are discussed and comparison made with its porcelain analogues. A comprehensive survey is made of extant examples of Coade stone and those that bear a range of factory marks which are described over the production period from the earliest in the 1770s, through the partnership with John Sealy from 1772 to the latest mark in 1837.

Keywords Lithodipyra · Coade · Coade & Sealy · Croggon · Marks on Coade stone · High-fired ceramics · Garden and building ornaments

1.1 The Phenomenon That Is Coade Stone

Historically, from the late eighteenth century into the fourth decade of the nineteenth century, for almost 70 years of continuous production, an artificial stone substitute for natural stone occupied a prime position in architectural work in Georgian and Regency Britain. This saw it reach an ascendancy for its use in building work and in the adoption of neo-classical designs and artefacts that addressed the desirability of wealthy clients, who engaged with architects and landscape gardeners to restructure their properties in Georgian Britain and overseas. The basic tenet of this artificial stone is that it was a moist paste which could be moulded into any shape or form and after firing treatment in a kiln for up to a week or so at a high maximum temperature of 1150 °C it became a hard fired ceramic material that was non-porous and extremely resistant to environmental degradation, unlike the natural stone it was designed to replicate. In contrast to the sculpted stone and marble it was designed to imitate, the artificial stone moulds could be used repeatedly to produce identical statues or building artefacts and intricate surface designs could be effected on the moist moulded paste before its firing process commenced to accommodate the personal wishes of the client, such as the insertion of an heraldic coat-of-arms or a crest, which was then finding vogue in society with the advent of armorial porcelain and armorial silverware for entitled armigerous families. Another advantage of this artificial stone substitute was that it could be made significantly more cheaply than

H. G. M. Edwards, C. J. Brooke, *Coade Stone*, Cultural Heritage Science,
https://doi.org/10.1007/978-3-031-71432-0_1

its corresponding analogues in marble, lead or bronze, so the neo-classical garden statuary which was then finding support for formal gardens as a result of travellers returning from the Grand Tour to Greece, Italy and Egypt was much more accessible and did not require the skills of expert sculptors and stonemasons to manufacture each piece individually to every commission.

The prime mover in the eighteenth century for artificial stone artefacts from 1769 was Eleanor Coade, who gave her name to her particular product as the eponymous "Coade stone", or slightly earlier even as "Coade's Lithodipyra", and she cornered the upsurge in the artificial stone market in neo-classical statuary with her astute business acumen and shrewd appreciation of the needs of her society clients. Unlike many of her contemporary porcelain and ceramics factory proprietors, Eleanor Coade did not come from a family with an established ceramics history or with a relevant background expertise and she had no personal experience of ceramics manufacture, her business experience rather was in the field of drapery and linen textiles and she and her mother ran their merchant textile shop in London which she acquired some years earlier in 1766. However, she saw a business opportunity avail itself when Daniel Pincot, the proprietor of an ailing artificial stone manufactory in King's Stairs, Lambeth, fell upon hard times and he sold his company to her for £4200 in 1769. For a few months, Eleanor retained the services of Daniel Pincot, but it is stressed not as a partner, although he publicly claimed rather irresponsibly to be the chief proprietor of the new firm, much to her annoyance, and early in 1771 she dismissed him from her business for blatant misrepresentation and dishonesty. It appears that in 1771 Daniel Pincot had submitted a copy of the classical Borghese Vase in artificial stone for an exhibition at the Society of Artists, which was described as "a very fine performance". He advised potential customers that copies of the Borghese Vase, along with the analogous Medici Vase, would be available for purchase from him: a Borghese Vase was ordered by Henry "The Magnificent" Hoare for his Temple of Flora at Stourhead House, and a pair of Borghese and Medici Vases for Kedleston Hall by Sir Nathaniel Curzon, Baron Scarsdale. Clearly, these transactions were accomplished without the knowledge of Eleanor Coade, who despite the prestigious clients involved, was highly displeased at Daniel Pincot's assumption of proprietorship in her company. On the 11th September 1771, an irate Eleanor Coade placed the following notice in the *Daily Advertiser*:

> Whereas Mr Daniel Pincot has been represented as a Partner in the Manufactory which has been conducted by him; Eleanor Coade, the real Proprietor, finds it proper to inform the Publick that the said Mr Pincot has no Proprietary in this Affair, and no Contracts or Agreements, Purchases or Receipts will be allowed by her unless signed or attested to by herself. Customers are told to contact her directly at Mr Demar's, Bridge Head, Lambeth.

Three days later, Eleanor Coade placed another notice in the *Daily Advertiser* advising readers that Daniel Pincot had no further employment at her manufactory at King's Arms Stairs and that all future orders should be directed to her there. Daniel Pincot now disappears from the scene and dies in 1792 (*The Gardens Trust*, 2016). It is interesting that Eleanor Coade did keep the Pincot moulds for the Borghese and Medici vases and reproduced them to order well into the 1800s.

Eleanor Coade then employed John Sealy, her cousin, as a partner and they together ran the firm of Coade & Sealy until John Sealy died in 1813, when she then took on the services of another relative into the business, William Croggon, until Eleanor died in Camberwell in 1821 at the age of 88. William Croggon ran the Coade artificial stone business of Coade & Co. thereafter until 1834, but he ran into financial difficulties and became bankrupt, before dying in 1835. His son, Thomas Croggon, did try to keep the firm of Coade & Co. going for a while but eventually it was sold off in 1838 and the moulds and designs were finally dispersed to competitors and successors at auction by Routledge & Co. in 1843—and the names of John Blashfield, Mark Blanchard and Felix Austin will surface later in this context.

The remarkable fact is that neither Eleanor Coade nor Daniel Pincot were the first to manufacture artificial stone, which had been known for some centuries hitherto, and although Pincot was clearly not successful in his manufacturing venture, other earlier manufactories also did not survive for any length of time to make it a profitable business. Pincot never patented his recipe and formulation for an artificial stone, which it is maintained historically had been first achieved in 1722 by Richard Holt, whose recipe although a secret and jealously guarded at that time, eventually revealed his stone composition publicly in 1730 in a legal deposition set up to thwart efforts by unscrupulous competitors who were intent upon subverting his patent. Pincot's recipe, in contrast, did not seem to be a secret but it was never publicised and it seems that it was certainly different in composition from that published by Holt. Although not formally publicised by Eleanor Coade, her formulation and recipe were later published openly in 1850 by a relative, Charles Fowler (Fowler, 1850). A little known but accurate recipe formulation for Coade stone had in fact been published earlier in 1821 by David Laing, architect to the Customs House in London, wherein the base raw materials of ball clay, fine quartz, crushed soda, lime glass, crushed flint and grog were divulged without revealing the proportions (Stanford, 2016):

> A species of terracotta. It combines in one mass, pipe-clay, flint, sand, glass and stoneware, that has already passed the furnace. These are ground to very fine powder, and are mixed together in the proper proportions, and the whole is well kneaded together by means of the addition of water. In this state it forms a kind of paste which has the ductility of the clay usually employed in modelling.

Historians have made several allegations that are thus far unfounded in their efforts to explain Eleanor Coade's successful manufacturing operation, including proposing the idea that she altered Daniel Pincot's recipe significantly upon taking over the firm: how Eleanor was able to effect such a major technical change in composition, alone and unaided, in such a short time and without any detailed practical knowledge of the ceramics industry or assistance from outside is not explained—in contrast, her contemporary ceramics manufacturers even with skilled help and expertise available on site could only achieve advances in their product quality over significant periods of time and after the institution of quite involved and lengthy experimental and empirical manufacturing trials. Other historians, whilst recognising this, have put forward the hypothesis that Eleanor Coade must have been scientifically adept and knowledgeable to understand the high-firing ceramics process and its use of selected raw materials but there is no evidence of this emergent from elsewhere. She was certainly an astute businesswoman and she rapidly transformed the ailing business of Daniel Pincot in 1769 by engaging with the finest sculptors and designers of her day such as John Bacon and Joseph Panzetta, and enlisting the support of the most eminent architects and designers such as Robert Adam, Sir Thomas Hope, Humphrey Repton, James and Samuel Wyatt, John Holland, John Nash, Sir William Chambers and Sir John Soane and of landscape gardeners such as Lancelot "Capability" Brown. Roger White has summarised the prevailing situation (in Kelly, *Mrs Coade's Stone*, 1990) that she afforded with her artificial stone artefacts:

> . . . allowing the great architects of the day to indulge in an orgy of classical ornamentation that would not have been possible in carved stone.

What has not been commented on previously is that Eleanor Coade chose not sell her artificial stone artefacts as job lots for knockdown prices at minor auction sales as did Daniel Pincot and the other contemporary manufacturers—who thereby did not recover anything like an economical sale for their manufactured artificial stone artefacts—but she set out to deal directly with the architects, gardeners and their clients, offering a wide portfolio of artefacts for standard commissions, which could be varied in size and artistic composition according to the wishes of the client. She rapidly gained the endorsement of Adam, Soane, Brown and upwards of 150 architects and designers as listed comprehensively by Alison Kelly (Kelly, *Mrs Coade's Stone,* 1990). What is equally surprising is that this upsurge in demand for artificial stone artefacts did not itself generate a flux of competitors to secure a slice of this burgeoning market—however, by producing practically the material that lived up to its advertised expectation of longevity of survival and engaging with the best people exclusively to apply it in their building and landscaping works commissions, Eleanor Coade had effectively shut out access for all her competition. Josiah Wedgwood, one of the most influential and successful English potters in Staffordshire in the eighteenth century, said in 1780 that he had been unable to break into the artificial stone ceramics market because Eleanor Coade had already secured the best architects and designers for her sole use and thereby had effectively denied him access to their skills and contacts for commissions.

1.2 Coade Stone Marks

Unlike its contemporary porcelain and other fired ceramic analogues, historians of ceramics will find it rather curious that the Coade artificial stone marks are not present upon many pieces and upon looking more closely, the marks used on Coade stone artefacts actually have a rather haphazard distribution and variability in their usage: Table 1.1 gives a compilation of the types of Coade stone marks that are found on artefacts manufactured between 1769 and 1840. One thing that is

Table 1.1 Impressed and dated marks used by Coade & Co., 1769–1837

Mark	Place	Artefact
COADE LONDON	Kedleston Hall, Derbyshire	Medici Vase[a]
COADE LONDON FECIT	Kedleston Hall, Derbyshire	Borghese Vase[a]
COADE'S LITHODIPYRA, LONDON 1787	Ruban, Ards	Keystone
COADE LONDON 1792	All Saints Church, Dewlish	Headstone of Charles Hill
COADE LAMBETH	Anglesey Abbey, Cambridgeshire	Caryatid
COADE LAMBETH 1793	West Dean College, Sussex	Statue of female on plinth
COADE LONDON 1793[b]	Rochester Cathedral	Monument to Lady Henniker
COADE LONDON 1793	Croome Park, Worcestershire	Druid
COADE LONDON 1797	Croome Park, Worcestershire	Keystone
COADE & SEALY 1800	Ham House	River God Statue
COADE & SEALY 1800	Buscot Park, Oxfordshire	Egyptian statue
COADE & SEALY LAMBETH 1800	Ham House	Pineapple
COADE & SEALY LAMBETH 1800	Buscot Park	Egyptian figure
COADE & SEALY 1802	St Mary's Church, Paddington	Monument to Joseph Johnson
COADE & SEALY 1804	Taplow Court	Urn and Statue of George III
COADE & SEALY LAMBETH 1805	Buckingham	Coat-of-arms Marquess of Buckingham
COADE & SEALY 1806	Wyvern Barracks, Exeter	Royal Coat-of-arms
COADE & SEALY 1806	Bicton Park	Busts of Nelson, Wellington
COADE & SEALY 1808	Butter Market, Chichester	Frieze
COADE & SEALY 1809	Bargate, Southampton	Statue of King George III
COADE & SEALY 1810	Lincoln Castle (Dunston Pillar)	Statue of King George III
COADE & SEALY 1810	Harlaxton Manor, Lincolnshire	Lion with coat-of-arms
COADE & SEALY 1812	Chatham Historic Dockyard	Royal Coat-of-arms
COADE & SEALY 1813	St Michael's Church, Bloxholm	Coat-of-arms of General Manners
COADE LONDON 1814	Sezincote	Indian bull figure
COADE 1815	Imperial War Museum	Royal Coat-of-arms

(continued)

Table 1.1 (continued)

Mark	Place	Artefact
CROGGON, LATE COADE LONDON 1822	Easton Neston, Towcester	Ornamental gateway[d]
CROGGON LONDON 1824	Shugborough Hall	Coat-of-arms of Earl Lichfield
CROGGON LAMBETH 1824	—	Lion
CROGGON LAMBETH 1826	Kew Gardens	Medici vase/ex Royal Lodge Gardens, Windsor
WWF Coades Sculpt[r] 24 May 1837	Westminster Bridge	South Bank Lion
T. CROGGON, LAMBETH	Suffolk Street, Westminster	Royal Coat-of-arms
COADE LAMBETH 1840	?[c]	Female statue on plinth

[a]Although undated, included here as house records indicate purchase in 1771—one of the earliest marked Coade stone artefacts
[b]Sometimes the marks have a comma after COADE and a full stop after LONDON. The mark may also occupy a single line or run over two or three lines
[c]Advertised for sale and provenance unknown
[d]in Alison Kelly (*Mrs Coade's Stone*, 1990) the mark is recorded as Croggan late Coade 1822 with an incorrect spelling of Croggon. However, van Lemmen (*Coade Stone*, 2006) shows a picture of a similar mark on the same structure which is correctly spelled!

immediately striking is the variability of the Coade factory mark, which seems to have little correlation with the year of manufacture, even when accompanied by a date! In the field of ceramics, factory marks and their modifications can usually be correlated with specific periods of manufacture, but this does not occur with Eleanor Coade's products, many of which seem not to bear any visible factory mark at all. From the data in Table 1.1. we can make several statements about these marks which may assist in the placement of future marked artefact discoveries into somewhere near the correct relative chronological period:

- Of the 33 dated marks from the Coade factory listed in Table 1.1 the earliest refers to COADES LITHODIPYRA LONDON 1787, closely followed by COADE LONDON 1792. Here, the earliest Coade mark on an artefact seems to be that of COADE LONDON (Fecit) but this is undated: it has been included in this list as supporting documentation refers to the purchase of the Medici and Borghese Vases at Kedleston Hall in 1771 which bear this mark, and which must therefore be some of the earliest neo-classical artefacts manufactured by the then fledgling Coade & Co. of Lambeth. Some of the latest marks are the marks on the underside of the paw of the South Bank Lion (The Lion Brewery lion), namely WFW Sculpt[r] COADES 24 May 1837, the unique T.CROGGON LAMBETH, which is undated, and the rather curious dated mark COADE LAMBETH 1840. The former script mark refers to the sculptor of the lion mould, William Frederick Woodington, and the date of installation of the statue at the Lion Brewery in Lambeth. William Croggon had already died in 1835 and the Coade business had then passed into the hands of his son, Thomas Croggon, who clearly had recognised the earlier Coade mark with a precise date. The final mark,

T. CROGGON LAMBETH, although undated, must post-date the inscription on the lion statue as it accompanies a Coade stone plaque of the Royal coat-of-arms of Queen Victoria, who did not accede to the throne until June 20th, 1837, the earliest date when these could then have been adopted. The third mark highlighted here of COADE LAMBETH 1840 on a female garden statue of an unknown provenance is especially interesting and it is quite possible that this is, therefore, the last definitively dateable piece of Coade stone as the manufactory closed down in 1838 and the lots were auctioned off in 1840. Although it may well have been purchased at the disposal of the factory's assets in 1840, the impressed date mark implies that it must be of a very late manufacture indeed, and the question could be posed as to who was around in the manufactory with the requisite skills in the Coade business to make such an artefact at that time? It is of course possible that an unmarked artefact was purchased at auction and the mark applied at that time.

- There are, thus, twelve distinct marks used by the manufactory as identified in this listing, embracing the 69 years of production between 1771 and 1840: although, the earliest stamped and dated mark is from 1787. An example of an early Coade mark is shown in Fig. 1.1., being COADE, LONDON. 1793, on a Coade stone monument to Lady Henniker in Rochester Cathedral, which depicted a lady (Truth) admonishing Father Time. This featured in an article in the *Gentlemen's Magazine* of 1794, Part 1, page 410, and the designer has historically been attributed to Thomas Banks, which has been disputed rather more recently.
- Apart from the two early rather solitary marks dated 1787 advertising COADE LITHODIPYRA, COADE LONDON seems to be the earliest established manufactory mark with a year date that was used consistently from about 1793 until 1797 (1771 if we accept the Kedleston Hall vases in this context although the date does not appear on the COADE LONDON stamp thereon), and thereafter COADE & SEALY with a date which appeared from 1800 until John Sealy died in 1813, when COADE LONDON reappeared again up to 1814. An example

Fig. 1.1 Early Coade stone impressed mark, COADE, LONDON. 1793, on the base of a memorial to Lady Henniker in Rochester Cathedral. Strictly referred to as Dame Anne Major, daughter of Sir John Major Bt of Worlingworth Hall, Suffolk, she died on the 18th July 1792, as wife of Sir John Henniker of Newton Hall and Stratford in Essex. The design has been attributed historically to Thomas Banks RA, but this has been disputed as Banks is not known to have worked with Eleanor Coade. The tomb is flanked by figures of Truth and Father Time, which have been attributed diversely to both Coade Stone and marble: despite this, the figure of Father Time has the impressed mark shown here!

Fig. 1.2 Impressed mark, COADE & SEALY. LONDON 1813, on the base of the tomb of Major General Bowes in Beverley Minster, Yorkshire. John Sealy died in 1813. General Barnard Foord Bowes served with Wellington in the Peninsular War and was severely wounded at Badajoz. He was killed in 1812 in storming the walls of Salamanca, leading his fusilier brigade comprising the 2nd, 32nd and 36th Regiments of Foot in the 4th Division. Although bearing the mark of Coade & Sealy, details of the commission from Mrs. Bowes appear in William Croggon's work book, costing 150 guineas for the memorial and 26 guineas for the packaging and transportation. Kelly (*Mrs Coade's Stone*, 1990, p. 251) has noted William Croggon's error in the Latin inscription on the Coade stone angel holding the honour roll: PRO PATRIAE

of the COADE & SEALY mark is shown in Fig. 1.2. An alternative mark, COADE LAMBETH 1793, has also surfaced on two female statues at West Dean College, Sussex (Ponting, 2003), which again was used intermittently until 1840, after the official closure of the factory in 1837. However, an alternative mark of COADE & SEALY LAMBETH also appears as a variant alongside these in the earlier years. There was, therefore, a considerable time lag operating between John Sealy joining Eleanor Coade in 1771 and the recognition of his key role in the company through the adoption of his name on the factory mark; this is rather curious as the first Catalogue of their designs of statuary and masonry embellishment for buildings appeared in 1784 and its title clearly is identified there as reproduced in Appendix II and is referenced as follows: E. Coade, "*A Descriptive Catalogue Of Coade's Artificial Stone Manufactory, At King's Arms Stairs, Narrow-Wall, Lambeth: Opposite White-hall Stairs. With Prices Affixed. Consisting Of, I. Statues. II. Busts. III. Vases. IV. Pedestals. V. Capitals. VI. Frizes. VII. Fascias. VIII. Panels and Tablets. IX. Medallions. X. Pateras for Elevations from 30 to 40 Feet. XI. Ditto from 20 to 30 Feet. XII. Pateras for Elevations from 10 to 20 Feet. XIII. Ditto of 10 Feet and under. XIV. Coats of Arms. XV. Imposts. XVI. Key-Stones. XVII.*

Architraves. XVIII. Rustics. XIX. Pinnacles. XX. Balusters. XXI. Balls and Feet. XXII. Chimney Tops. XXIII. Piers. XXIV. Consoles. XXV. Sepulchral Ornaments. XXVI. Chimney-Pieces. XXVII. Pieces of Furniture and various ornaments. XXVIII. Ornaments to fix on Stone or Wood. XXIX. Wood Chimney-Pieces with Stone Ornaments. With an Appendix, 1784".

- A reversion to the mark of COADE LONDON and the year date occurs for 1814.
- Then COADE and the year date appears in 1815.
- A new type of mark appears when William Croggon purchases the manufactory after the death of Eleanor Coade in 1821, an example being CROGGAN late COADE 1822. It has been pointed out by Alison Kelly that the name of CROGGON has been mis-spelled in this mark! The only known occurrence of this mark to date is on the Coade stone ornamental gateway to Eaton Neston, Towcester, Northamptonshire, which now comprises the archway leading to Towcester racecourse, although there are several pieces of Coade stone in the structure, including a coat-of-arms, which seem to be unmarked. This mis-spelling, if truly recorded on the figure of a hind, is a feature of concern as any similar mistakes in analogous impressed ceramic porcelain marks of this period usually is indicative of a clear indication of a fake or forgery! However, van Lemmen reproduces this mark in his book (van Lemmen, *Coade Stone*, 2006, p. 6) which occurs on the base of a statue of a hind decorating the ornamental archway (Fig. 1.3) and it is there correctly spelled as CROGGON, so there may well be a more modern observational transcription error made elsewhere.

Fig. 1.3 Impressed mark, CROGGON LATE COADE LONDON. 1822, which he used after the death of Eleanor Coade in 1821, upon the base of a hind statue surmounting the roof of the lodge adjacent to the ornamental archway, called the Brentford Screen, to the Easton Neston Estate at Towcester Northamptonshire. This had been apparently incorrectly recorded by Esdaile (1940, 1946), and afterwards elsewhere by Kelly, as the mis-spelled CROGGAN LATE COADE 1822, which was later corrected by van Lemmen

- By 1824 William Croggon was using the mark CROGGON LONDON with the year date, and a year or so later CROGGON LAMBETH with the year date, dropping the allusion to COADE entirely (Fig. 1.4).
- A latter type of dated mark associated with Coade stone appears on the largest artefact manufactured in Coade stone, namely the lion made for James Goding's Lion Brewery, Lambeth, which has the legend WFW Sculpt^{r} COADES 24 May 1837 written under one paw (Fig. 1.5). This is an unusual mark in several respects

Fig. 1.4 Impressed mark, CROGGON LAMBETH 1826, on the base of a Medici Vase in Kew Gardens. The Medici Vase, along with the Borghese Vase, were originally commissioned in 1825 by John Nash for the Royal Lodge, Windsor, where they sat atop a specially designed pair of pedestals. The bill for the vases and their pedestals came to £293-6s-8d. This Borghese vase is now in Wrest Park, Bedfordshire

Fig. 1.5 Unusual and unique script mark printed on the paw of the South Bank Lion, *WFW Sculptr COADES 24th May 1837,* which was only noted when the statue was removed from its original location atop the Goding Lion Brewery in Lambeth in 1949 preparatory to site demolition to clear space for the Festival of Britain in 1951. Currently, this is the last known definitively dated mark on a certified piece of Coade stone sculpture although others of a later date are noted in the text

as, firstly, it cannot be seen by an observer of the statue when in position and was only noted when the lion was lifted from its plinth during its relocation to the Waterloo station concourse in 1949 and removed thence to a plinth on Westminster Bridge in 1966, where it still stands and is known now as the "South Bank Lion". The mark refers to William Frederick Woodington, the sculptor, who became President of the Royal Academy of Sculptors in 1851, and, secondly, it unusually bears the day and date of its manufacture. A second, smaller lion from the Lion Brewery now graces the Rowland Hill memorial gate at Twickenham stadium, where it was gilded to celebrate the hosting of the Rugby World Cup there in 1991, being representative it is said of the British Lions rugby touring team; it is not known whether this lion bears a hidden mark. A third lion from the original Lion Brewery was lost upon the demolition of the site in 1949 for the Festival of Britain. The Lion Brewery nominated business has since been reconstituted afresh at a new base in Singapore but still retains the lion logo on its beers. On a less certain note, a three—ton wooden copy of the South Bank Lion, painted red, and said to be tentatively attributed to its designer is now displayed at the Cambridge University Rugby Union Football Club ground at Grange Road, Cambridge, having spent some time at the Lion Yard shopping centre in Cambridge. Further discussion relating to these three lions will be undertaken later under the Case Studies section to include some more information from recent research discoveries.

It is not immediately obvious if the impressed mark changed with the size of artefact manufactured as happened with several contemporary porcelain manufactories: it is suspected that this did not occur, as in most cases the artefacts constructed in artificial stone are significantly larger in size anyway compared with their fine china contemporary analogues. The placement of the impressed mark on the Coade stone artefact is also variable but in most cases it is readily visible to an observer with the exception of the South Bank Lion, the heaviest piece of sculpture made by the Coade manufactory, where the mark is placed underneath the paw of the lion and could only be seen when the artefact was lifted off its plinth prior to its relocation. In several cases where Coade stone pieces are now being offered for sale at auction, the presence of marks which may have been hidden by their mounting in their original locations is noted, such as the undersides of paterae or heraldic tablets featuring coats-of-arms: these have not been included in our specific survey here as they are generally undated.

The frequency of occurrence of the marks on Coade stone artefacts is not immediately obvious from the above analysis as the data refer specifically to the most commonly cited artefacts in the literature. For this, it is necessary to consider a wider range of artefacts and this requires a complete listing of every Coade stone artefact to be undertaken for a meaningful analysis to be made.

1.3 Analysis of Coade Stone Artefacts and Their Comparison with Porcelain

Although both Coade stone and porcelains are high temperature ceramic materials that have been synthesised from similar types of raw materials that include clays, sand, grog and glass, the marks on porcelain artefacts are seemingly considered to be much more important by connoisseurs for their attribution than are those of Coade stone and hence have been studied in more depth. The assignment of an unmarked artefact to the category of Coade stone historically is, therefore, very much still the opinion of the observer, which is only forensically sustainable when supported by historical documentary evidence, which is frequently sadly lacking. Several instances will be recounted later where errors in attribution of an artefact to Coade stone seem to have occurred, especially where the artefact is not readily accessible and may be placed rather inaccessibly high on a building. Whereas porcelain manufactory marks have been observed to change predictably with specific periods of production, those of Coade stone have generally only received a casual mention in the literature and none of the previous literature has attempted to analyse these marks comprehensively. In many cases, it appears that the dating of a Coade stone artefact is accomplished through details that have been preserved in existing archival documentation, such as the *Handbook of Coade's Gallery* of 1799, which listed their full repertoire of artefacts, and up until that time the Coade artefacts and commissions that had been undertaken and listed from notes made of Eleanor Coade's memory recall, which is not always a completely accurate record as Alison Kelly (*Mrs Coade's Stone*, 1990) has ascertained in her research. Other documentary sources are the invoices for work which are preserved in the specific house records of the time, such as the purchase of the Coade stone sundial at Weston Park and the Medici Vase at Castle Howard. Another source of Coade stone commissions is the workbook of William Croggon (now held in the Public Record Office, London), who joined Eleanor Coade after John Sealy had died in 1813 and he maintained a record of work that he was involved with at the company between 1813 and 1821, after which Eleanor Coade died and when he took over the business operations completely, for which he had to submit details to the Chancery Court to facilitate his purchase of the Coade & Co. business which Eleanor had bequeathed to charitable causes. After 1821, William Croggon did maintain a partial list of invoices and correspondence with clients which has also proved useful for attribution purposes, although Alison Kelly (Kelly, *Mrs Coade's Stone*, 1990) did find that she was unable to correlate some of this correspondence with the current presence of Coade stone artefacts at their same locations at the specified site. A further source of Coade stone commissions relates to the papers of Sir John Soane (1753–1837), an eminent architect who used Coade stone artefacts in his work for 48 years from the age of 27 in 1782 until 1830. An avid collector of classical antiquities, he was very much in favour of Eleanor Coade's neo-classical designs and he used them extensively in many of his architectural assignments including his own country house at Pitzhanger Manor as well as his London residence at 13, Lincoln's Inn Fields. Soane used

Coade stone artefacts in some of his very earliest work, such as the frieze which was later numbered 186 in the Coade and Sealy Catalogue in his commission at Adam's Place, Southwark, in 1780. Generally, then, although the assignment of a date to a particular Coade stone artefact accompanies the listing in a gazeteer such as that given in the books of Alison Kelly (*Mrs Coade's Stone*, 1990) and Hans van Lemmen (*Coade Stone*, 2006), the impressed marks are not always provided—which is a shame as these are usually also dated, except for the special one, for example, of COADE & CO., which is used undated. In such cases, dates for the commission have then been provided additionally from the documentary sources listed above.

1.3.1 Numerical Analysis of Registered Coade Stone Artefacts

In this section, the locations and descriptions of Coade stone artefacts as listed in the comprehensive gazeteer compiled by Alison Kelly (*Mrs Coade's Stone*, 1990) have been analysed: in her compilation, Kelly has drawn on several local historical works which have described the significant buildings and of their possession of Coade stone artefacts, which rather means that there is often a subjective decision made by observers about their attribution, which Kelly has acknowledged but had not been able to confirm in several instances, and in these cases the authors here have listed these in the "questionable attribution" category. In many cases, too, the artefacts that were originally installed in their specific locations have suffered a potential relocation elsewhere arising from the demolition of the original buildings concerned, or from their acquisition at auction sales of the dispersal of estates, or from damage caused in the bombing of London particularly between 1940 and 1944. In several cases, the number of artefacts recorded as having been supplied to a house has been reduced upon later inspection through being sold off or damaged and removed. Finally, in other cases, we should be aware that the observation of a piece of sculpture should not imply that it is definitively a Coade stone artefact when an observer expects to see it there in that location and in several instances there is a subsidiary documentary record that earlier sculptures have been replaced at a later date, sometimes by Coade stone replicas, but in other instances not so, and the *vice versa* situation also applies. The shape and style of an artefact should also not give one reason to assume it is typically Coade stone and on two cited occasions an expert observer has revealed that the expected artefact of Coade stone is actually a replacement sculpture made of Portland stone *in situ*, which has been effected for some unknown reason! For example, it was noted that a genuine Coade stone plaque and paterae originally commissioned by Lord Rossmore in 1800 for Monaghan Market Hall, Co. Monaghan, Ireland, have been replaced later (Brett, *Court Houses and Market Houses of the Province of Ulster*, 1973) by Portland stone sculptures of exactly the same design. This not a unique occurrence and a second noted instance is provided by Lucan House, Dublin, where a Coade stone memorial to Patrick Sarsfield has been replaced by one of Portland stone (Dr. D.J. Griffin cited in

Kelly, *Mrs Coade's Stone*, p. 438, 1990) despite its appearance in a genuine Coade stone form elsewhere. This scenario, therefore, will assume greater significance when we come to assess the suitability of selecting Coade stone specimens for future chemical analysis and the establishment of a Coade stone database, where the determination of their composition will only be definitive if the specimen has been verified as being an absolutely genuine Coade stone exemplar in every respect. The restoration and conservation of damaged Coade stone artefacts is a different matter and for other reasons these can compromise the subsequent analytical interpretation, in that the presence of resins, new modern composites and adhesives will undoubtedly skew an analytical determination whereby an apparently genuine Coade stone figure is registered as indeterminate or not matching in its composition with the analytical standard database. A perfect example of this situation is provided by the statue of King George III which was erected originally on the Dunston Pillar in Lincolnshire to celebrate his 50 years on the throne in 1810: this was demolished in 1942 as a hazard to low-flying aircraft and the bust was reconstituted from several dozen fragments that now stands in the grounds of Lincoln Castle. It will transpire that elemental analysis of the King George III bust definitively decrees that it is apparently not of Coade stone, a result that is ascribed to the complex composition of modern resins, adhesives and formulations used in its restoration which affect the chemical analysis.

The numerical analysis carried out on the comprehensive survey carried out by Alison Kelly on Coade stone artefacts and published in her scholarly text (*Mrs Coade's Stone*, 1990) has been effected here and the results are as follows:

- Although Alison Kelly has been cited as identifying almost 700 Coade stone sites in her comprehensive gazeteer, a numerical analysis reveals that the number of pieces involved is considerably more because many of these sites have multiple examples of Coade stone artefacts: it is difficult to quantify exactly the number of pieces involved as in some cases the multiplicity of specimens is very large, such as the provision of balusters, plaques and paterae which could easily reach double figures and more at any particular site, and the identification of whole streets of houses at one site, which could number many houses displaying an indefinite proliferation of keystones, voussoirs and imposts made of Coade stone. Examples of the latter scenario are Harley Street, London, where 40 houses between Nos. 66 and 145 may each have more than 16 different combinations of Coade stone on their doorways, Wimpole Street, London, where 16 houses between Nos. 19 and 56 have Coade stone decoration, and Bedford Square in Bloomsbury, London, where an unspecified number of houses have Coade stone decoration applied to their walls and doorways. The Livery Company Halls in the City of London also display a varying number of Coade stone features: in all, some 13 Livery Companies commissioned work from Eleanor Coade and her successors, namely, The Skinner's, Watermen's, Cordwainers', Grocers, Vintners, Armourers, Waxchandlers, Stationers, Merchant Taylors, Mercers, Carpenters, Leathersellers and Fishmongers. Much Coade stone embellishment to these buildings has been lost during their rebuilding, except for those of the Watermen's and Vintners Companies. A classic exemplar of a single Coade stone site that contains many exemplars of architectural embellishment is that of Lord Schomberg's house in Pall Mall, London, which was built in 1690 and was embellished in 1790 with multiple additions of Coade stone to its facings and also then acquired a Coade stone *porte cochere*, which displays some superbly preserved statuary, pillars and pediments.

- The numbers of specimens of Coade stone that have been distilled from the list of sites given by Alison Kelly has been summarised here in Table 1.2, and have been described under the headings as were mentioned in Eleanor Coade's specific list (CG, 1799), documentary evidence supplied from the workbooks of William Croggon (CO and CD, 1813–1837) and statements provided in local historical tracts by other historians and chroniclers (termed Other). In addition, the number of Coade stone artefacts that were no longer extant or whose presence could not be verified directly in observation by Alison Kelly and others, and the number which could be deemed to have been destroyed by the demolition of the houses and estates concerned, are tabulated for each section. There are four sections in Table 1.2, comprising sites in England (outside London), London, Wales, Scotland and Ireland together, and Overseas. A final column in the table refers to the identified specimens of Coade stone that have been painted or bronzed, which often then obscures their origin (and potential identification!), and also to specific items of *scagliola* that were supplied by Coade & Co., especially after William Croggon took over and diversified the business after 1821, often along with their Coade stone products. In the second decade of the nineteenth century, the business started to diversify into the production of *scagliola* to reflect the changing tastes of Georgian society, who wanted examples of this wondrous

Table 1.2 Numerical Analysis of Coade Stone Specimens from Kelly Lists[a]

Eleanor Coade CG, 1799	William Croggon CO, CD, 1813–1837	Others	Questionable attribution[b]	Destroyed/ Estate demolished[c]	Painted/Bronzed & Scagliola	
					P	S
1. England (except London)						
166	137	310	135	111	17	10
2. London						
61	137	225	66	96	16	27
3. Wales, Scotland and Ireland						
55	36	82	26	38	1	4
4.Overseas						
8	18	17	13	3	0	0

[a]Alison Kelly, *Mrs Coade's Stone,* 1990: Gazeteer, pp. 331–445
[b]This category covers the disappearance by 1990 of the Coade stone specimen from the designated site where its presence had been observed or recorded hitherto: in some cases an acquisition was made and the specimen was transferred elsewhere, thereby losing its original provenance unless recorded
[c]The destruction or demolition of the buildings or estate housing the Coade stone artefact includes vandalism and wartime bombing, especially in London and other major cities

new variant of coloured Italian *faux-marble* composite for furniture, floor and wall decoration.

- Occasionally, Coade stone relics have been noted in unusual locations for which their original purpose was not designed: an exemplar is provided by the four standing pillars in the Botanic Gardens at Portobello overlooking the Firth of Forth in Edinburgh. These have been described by local historians, who have queried their purpose there, but their characteristic shapes mark them out to be Coade stone chimney pots, probably recovered from a now derelict mansion, as compared with those that are still functioning for that purpose, for example, in Dalmeny House, Edinburgh.
- A summary of the data reproduced in Table 1.2 provides the following interesting information: in all, some 1252 Coade stone specimens have been identified by the authors from the 680 sites specifically listed in Alison Kelly's gazeteer—this does not include the indefinite multiplicity of streets or columns or balustrades provided in invoices or which are still identifiable at various sites but it does include individual and specified items, such as a Borghese Vase and a Medici vase supplied to the same site. London represents the location of some 30% of surviving Coade stone artefacts compared with the rest of the British Isles and Overseas. The actual data are as follows: British Isles excluding London 67%, London 30%, Overseas 3%. Some 15% of the Coade stone artefacts that were specified as having been manufactured for different sites have now disappeared or cannot be verified anymore—this does not include the 34 painted, gilded or bronzed items noted (2%) for which it is particularly difficult to identify the substrate visually without removal of the superficial paint, gilding or bronzing; it is quite possible that several of these are actually replacement composite items for an original Coade stone artefact. The inclusion of the 41 *scagliola* items in the Table is based upon the increasing popularity of this medium in the early nineteenth century, representing 3% of the total number of Coade stone artefacts supplied by Coade & Co during its existence from 1769 to 1837. Most of the *scagliola* was supplied to London-based clients, not forgetting that many of these would also have had their country estates and retreats, such as Sir John Soane (Pitzhanger Manor) and the Duke of Northumberland (Alnwick Castle), and the artefact, although ordered by a London-based client, may well have been intended for such a country seat rather than the London *pied-a-terre*.
- A second book on Coade stone appeared 16 years after that of Alison Kelly by Hans van Lemmen (*Coade Stone*, 2006): this booklet monograph gives the salient features of Coade stone but it is not as comprehensive a survey as that of Alison Kelly and gives only some 82 locations (70 in England including London, 2 in Wales, 6 in Scotland and 4 in Ireland, with 0 overseas). It is noteworthy, however, for its colour images of several artefacts and does display a number of Coade stone marks, which unfortunately is nowhere nearly representative of the large number of variations employed and therefore perhaps may give a false impression to the reader of the range and type of mark employed by Eleanor Coade and her successors. Also, more seriously as regards the content current text, the composition, raw materials and analytical studies applied to Coade stone ceramics are

not even mentioned. All 82 sites specified in the van Lemmen text are replicates of those presented in the more comprehensive study of Alison Kelly and no new discoveries or clarification of existing sites are mentioned therein.

1.3.2 Survey and Analysis of Coade Stone Manufactory Marks

In Table 1.3. is given an analysis of the full complement of Coade stone specimens which depict an impressed mark (and one example of a written mark) of a type, albeit sometimes a damaged or incomplete mark: the purpose underlying this study is to complete a narrative of all known Coade stone marks and dates supplied therewith to facilitate an investigation into the periods when these marks were used. It will transpire that, unlike its contemporary porcelain manufactories, the Coade manufacturing business did not seem to set a great store upon marking its pieces, perhaps because there was so little competition, if any, from erstwhile competitors and forgers who bedevilled the production of the corresponding early porcelain industries such that marks were considered a vital part of their ceramic production. Even so, some care needs to be taken in the acceptance of marks on porcelain and fine china as these, particularly the enamelled script marks, which were frequently copied by forgers and others who wished to pass off their wares as the genuine article: the firm of Samson et Cie. of Paris in the mid-nineteenth century deliberately applied fake script marks to their hard paste porcelain copies of earlier more desirable factories, but generally experienced collectors and connoisseurs can detect fundamental flaws in the overdecoration of these pieces—and, of course, the fact that a hard paste porcelain substrate as used by Samson et Cie. was invariably used to copy a soft paste porcelain original! The example of Chinese porcelains can be cited in this context, where the seal marks of earlier Emperors in cobalt blue pigment were frequently applied to later wares as a sign of "authenticity" for potential purchasers, such as the Ming Dynasty seal marks (1360–1643) that were later applied to Kangxi period (1660–1720) porcelains which were fired in the same Jingdezhen Dragon kilns using the same raw materials over the centuries. To English and Welsh porcelain connoisseurs, the incorrect depiction of a factory mark will cause alarm bells to ring concerning the attribution and potential authenticity of an otherwise apparently perfectly acceptable piece of porcelain, and the omission, inclusion or misplacement of a comma, hyphen or full stop, the use of incorrect capitals and overlarge sizes, will result in a fake classification being automatically bestowed upon the ceramic piece under evaluation by a connoisseur. A classic example of this is a piece of fake Nantgarw china that has the incorrect impressed mark of NANTGARW CW, which should read NANT-GARW C.W., and this would result in the outright and clear rejection of the piece in question as being genuine by a connoisseur. The placement of punctuation in the impressed Coade stone marks, however, seemingly cannot be relied upon for an equally definitive assignment or

Table 1.3 Coade Stone Specimens Bearing Impressed Marks and Occasionally Dated

Site/Estate	Artefact Description	Impressed Mark / Date	Comment
St Mary's Church, Buckland	Memorial to Elizabeth Perfect	COADE & SEALY 1802	
Buscot Park	Egyptian Pharaoh Figures	COADE & SEALY 1800	
All Saints' Church, Colehill	Memorial to Mark Pleydell	COADE & SEELEY 1802	Incorrect SEALY
Buckingham	Bridge plaque: Coat-of-Arms	COADE & SEALY 1805	
	Swan plaque	COADE & SEALY 1809	
Ritchings House	Classical frieze	COADE, LAMBETH 1792	
Taplow Court	King George III as Hadrian	COADE & SEALY 1804	
Anglesey Abbey	Caryatids	COADE, LAMBETH 1793	
Ollerton Lodge	Royal Coat-of-Arms	COADE & SEALY 1804	
St Michael's Church, Mevagissey	Memorial to Captain Bettesworth	COADE & SEALY 1812	
St Michael's Church, Lawhitton Calstock	Memorial to Richard Coffin	COADE & SEALY, LAMBETH 1796	
Bicton Park	Lord Nelson Bust	COADE & SEALY 1806	
	Sir Walter Raleigh Bust	COADE & SEALY 1806	
Wyvern Barracks, Exeter	Royal Coat-of-Arms	COADE & SEALY 1806	
Killerton, Exeter	Borghese Vases	COADE & SEALY 1805	
Princetown	Urns with POW Feathers	COADE, LONDON 1797	
All Saints Church, Dewlish	Headstone to Charles Hill	COADE, LONDON 1792	
Hill Hall, Epping	Column	COADE, LAMBETH 1791	
Bridge End Gardens, Saffron Walden	Borghese Vases	COADE & SEALY 1810	
Northwick Park, Moreton-in-the-Marsh	Lion plaque	CROGGON 1832	
Sezincote, Moreton-in-the-Marsh	Bulls, elephants, antique Apollos	COADE, LAMBETH	Some have dates 1813, 1814
	Borghese Vase, Suraya statue	COADE, LAMBETH	
Kineton House, Thornbury	Grecian sphinxes	COADE, LAMBETH 1840	Unusual Date

(continued)

Table 1.3 (continued)

Site/Estate	Artefact Description	Impressed Mark / Date	Comment
St James' Church, Bramley	Brocas Chapel monument	COADE & SEALY 1802	
Mottisfont Abbey	Urns	COADE 1794	
Somerley, Ringwood	Warwick Vase	COADE 1840	Unusual Date
Bargate, Southampton	King George III as Hadrian	COADE & SEALY 1809	
Swan PH, Rickmansworth	Inn sign	....OADE, LAMBETH	
Chatham Dockyard	Royal Coat-of-Arms	COADE & SEALY 1812	
Congregational Church, Faversham	Monument to John Simmonds	COADE & SEALY 1805	
Rochester Cathedral	Monument to Lady Henniker	COADE, LONDON 1793	
St Mary the Virgin, Speldhurst	Monument to William Nesbit	COADE 1807	
St Peter and St Paul's Church, Tonbridge	Monument to Henry Harpur	COADE & SEALY 1800	
Kirk Braddon, Isle of Man	Monument to Lord Henry Murray	COADE & SEALY 1800	
Coleorton Hall, Ashby-de-la-Zouch	Borghese Vase	CROGGON 1827	
		Medici Vase	
St Michael's Church, Bloxholm	Coat-of-Arms, General Manners	COADE & SEALY 1813	
Heydon Hall, Aylsham	Heraldic Panel	CROGGON 1830	
St Stephen's Church, Norwich	Monument to Elizabeth Coppin	COADE & SEALY 1812	
St Andrew Thorpe Church, Norwich	Monument to Elizabeth Meadows Martineau	COADE & SEALY	
Easton Neston	Archway	CROGGAN LATE COADE 1822	Mis-spelling CROGGON
Alnwick Castle	Medallion Duke & Duchess Northumberland	COADE	
Ponteland Church, Blyth	Monument to Richard Newton Ogle	COADE	
All Saints 'Church, Babworth	Monument to Mrs. Simpson	COADE & REPTON	Designed by Humphrey Repton
Barnby Manor,	Newark	Unicorn	COADE & SEALY
	Borghese Vase	COADE & SEALY	
Thoresby Park, Retford	Vase	COADE & SEALY 1802	

(continued)

Table 1.3 (continued)

Site/Estate	Artefact Description	Impressed Mark / Date	Comment
St Peter's College, Oxford	Plaque: Oxford Canal Co.	COADE 1790	
Blaise Castle	Borghese Vases	CROGGON	
Hutton St George Church	Coat-of-Arms	COADE & SEALY 1812	
Shugborough Hall	Coat-of-Arms: Anson	CROGAN 1834	should read CROGGON 1834
Gt Saxham Hall, Bury St Edmunds	The Umbrello garden building	COADE & SEALY	Umbrello: type of gazebo
St Mary's Church, Henstead, Beccles		Monument to George Mitchell	COADE & SEALY 1806
St Peter's Church, Chertsey	Aisle tracery	COADE & SEALY 1806	
Ham House, Richmond	Pineapples (12 off): 9 stamped	COADE & SEALY 1800	
	1 stamped	COADE & SEALY 1799	
	1 stamped	COADE	
Clerk House, Duchess St/Deepdene	Sphinx garden throne	COADE 1800	Designed by Thomas Hope
	Sphinx garden throne	COADE LAMBETH 1800	
Leith Hill Place, Dorking	Statues; light sconces	COADE 1797	
Runnymede Park, Egham	Plaques	COADE, LONDON 1791	
Brightling Park	Garden Seat	COADE SEALY 1803	& Missing
St Nicholas' Church, Brighton	Tomb of Anna Maria Crouch	COADE & SEALY	
	Urn	CROGGON	
Church Army Building, Brighton	Statues of reclining ladies	CROGGON 1823	
Market House, Chichester	Coat-of-Arms	COADE & SEALY 1807	
St Nicholas' Church, Old Shoreham	Monument, no name given	CROGGON 1828	
St Mary's Church, Petworth	Royal Coat-of-Arms	COADE & SEALY 1812	
Park Crescent, Worthing	Archway	CROGGON 1829	
All Saints' Church, Aston, Sheffield	Plaque to William Mason	COADE & SEALY 1804	
Hazlewood Castle, Tadcaster	Unicorn	COADE & SEALY 1801	

(continued)

Table 1.3 (continued)

Site/Estate	Artefact Description	Impressed Mark / Date	Comment
West Dean College, Sussex	Female statue	COADE LAMBETH 1793	
Hillam Hall	Dog Kennel	COADE & SEALY	
Harlaxton Manor	Lion	COADE & SEALY 1810	Ex-Witley Court
Culzean Castle, Ayr	Egyptian Lionesses	COADE, LAMBETH 1802	
Athenry Friary	Urn	COADE, LONDON 1790	Smoke, flames: resurrection
			Unique design
South Bank, Westminster	Lion: base of paw	WFW Sculpt^{r} COADES 24 May 1837	Script not impressed
6, Suffolk Street, Westminster	Royal Coat-of arms of Queen Victoria	T. CROGGON, LAMBETH	
Bolton Castle, Kildare	Coat-of-Arms: Earl of Kildare	COADE, LONDON 1795	
Dawson Grove, Monaghan	Keystone	COADE & SEALY	
Galtrim House, Trim, Meath	Lions	 1802	Obscured
Pakenham Castlepollard	Sphinxes	COADE'S LITHODIPYRA	
Rubane, Ards	Keystone, paterae	COADE'S LITHODIPYRA, LON-DON 1787	
Bank of Montreal, Canada	Plaques	COADE 1819	
Ryan Mansion, New York	Sphinxes	COADE'S LITHODIPYRA 1787	
The Capitol, Washington DC, USA	Frieze	1793	Obscured
Haiti	Lord Nelson Bust	COALE & LEALY 1816	Mis-spelling SEALY

not, and as we have noted above, the occasional misspelling of names such as Sealy, Lealy and Croggon in the mark are deemed to be still acceptable as genuine marks.

An analysis of the marks displayed on Coade stone specimens in Table 1.3 now follows:

- The 94 Coade stone artefacts currently known which depict a manufactory mark are listed in Table 1.3. The greatest majority of these are impressed marks, and only one artefact displays a painted mark, namely the lion statue created for the Lion Brewery in Lambeth in 1837, which has a painted script mark, *WFW Sculpt^{r} COADES 24 May 1837*, under its paw which was only revealed when the lion statue was lifted off its plinth and moved from its archway location when the Lion

Brewery was demolished in 1949. The WFW refers to the sculptor of the mould, William Frederick Woodington, who later became the eminent President of the Royal Academy of Sculpture in 1851. This is the only instance found so far where "COADE" has appeared in written script and the fact that it was used as late as 1837, after William Croggon had died some two years before, is perhaps significant in this respect. An even later mark has just surfaced, namely COADE LAMBETH 1840, on a garden figure with no known provenance, which nevertheless seems genuine.

- The use of 11 different factory marks is also unusual when we consider that only 94 artefacts from a total of almost 1300 discrete Coade stone specimens are actually listed as being marked—a surprisingly small number, and representing only 7% of the complement of manufactured surviving specimens. The first manufactory date impressed mark recorded appears in 1787, when we have two instances of the COADE'S LITHODIPYRA mark appearing. This means that Eleanor Coade had been running the manufactory for almost 20 years before this first mark appeared and we could well pose the rhetorical question … why did she feel the need to use a mark at all at that time but not for the preceding 18 years? The reason cannot just be advertising as by that time the business was well established and operating very successfully and the marking of only two items from the hundreds if not thousands that must have been created by that time seems an unrealistic target. A third COADE'S LITHODIPYRA mark is recorded but is undated but this surely must belong to this era also.
- The next earliest manufactory mark creates another complexity for our consideration since there were three marks being used simultaneously and interchangeably between 1791 and 1796: namely, COADE, LAMBETH; COADE, LONDON; and COADE.

  These marks disappear after 1796 but are revived again in the 1800s and COADE, LAMBETH and COADE, LONDON are even noted on artefacts that are reliably dated to about 1840, some years after the Coade factory had formally ceased production. The printed mark on the South Bank Lion also uses the designation COADES in a script mark from its manufacture in 1837!This could perhaps be a response to the ever growing presence of the perceived inferior cementitious alternatives that were then starting to appear on the market in competition with the Coade stone. It is said that Felix Austin of Austin & Seeley would exhort landowners to permit him to make his own moulds of their existing Coade stone figures so that he may cast his own versions of the same in an inferior cold-cured cementitious stone!
- The predominant manufactory mark is clearly COADE & SEALY, which appears on 50% of the known Coade artefacts, occupying the temporal range from 1799 to 1813, when John Sealy died. In the interim, it seems that the manufactory reverted to COADE, LAMBETH for 1813 and 1814, and there is even a badly mis-spelled version of COALE & SEELEY appearing in 1816! The same year a Coade stone statue of Lord Nelson bears a COADE & LEALY 1816 mis-spelled mark. The earliest mark involving John Sealy's name is COADE & SEALY,

LAMBETH which appeared first in 1796 some 25 years after he had joined the company.

- After 1813, when William Croggon replaced John Sealy at the manufactory, no change appears in the marks to incorporate his name and no new factory marks were used at all with the one exception of the mis-spelled COALE & SEELEY, which was used once in 1816, three years after John Sealy had died, which again is rather mysterious—the same can be implied for the COADE & LEALY 1816 mark which has been referred to above.
- The first mark to exhibit the name of William Croggon appears in 1822, just after Eleanor Coade had died and William Croggon had purchased the manufactory from her estate. In 1822, the sole mark of CROGGON LATE COADE appears on the gatehouse archway and lodges at Easton Neston in Towcester: there is some confusion about this mark as Katherine Esdaile and Alison Kelly have both recorded it with the mis-spelling of CROGGAN but Hans van Lemmen records it as being properly marked as CROGGON.
- From 1823 until 1834, William Croggon exerts his influence on the factory marks, using the name CROGGON, although his 1834 mark is also mis-spelled uniquely as CROGAN. His mark appears nine times in Table 1.3, in a third position only to COADE & SEALY and COADE LAMBETH.
- Occasionally, dates are found impressed into Coade stone artefacts without the accompanying mark: an example is that of the font, certified as being of Coade stone, in St James' Church, Chignall, dated 1838, without its accompanying factory mark.
- A very unusual mark appears, undated, as COADE & REPTON on a church memorial to Mrs. Simpson, wife of the local landowner, that had been designed by the architect Humphrey Repton. Despite the support given to Eleanor Coade by eminent architects and designers over many years, such as Robert Adam, John Bacon and Sir John Soane, this is the only example of a designer's name appearing on the actual factory mark alongside Coade. Although the memorial is undated, genealogical research indicates that in 1784 John Bridgeman married Henrietta Frances, daughter of Sir Thomas Worsley Bt of Appuldurcombe, who assumed the name of Simpson when he inherited Babworth Hall. Mrs. Simpson died in 1791, so the memorial would have been dated to around 1791–1793, and this would have come within the time frame of the COADE, LAMBETH and COADE, LONDON usage of factory marks. It seems that Humphrey Repton must have persuaded Eleanor Coade to incorporate his name uniquely at this time into her factory mark, perhaps when she was just embarking upon the usage of his designs for her products and needed to acquire and maintain his services.
- Of the 94 assigned factory marks listed in Table 1.3 we have identified six serious spelling errors, notwithstanding any other errors that may have been incorporated into the factory stamp by the omission of commas or fullstops. When one considers that the impression would have had to be made into the wet artificial stone paste from the initial moulds before being fired in the kiln as it would have been impossible to make an impression upon the hard-fired ceramic afterwards, the mistakes would have occurred in the stamps themselves or perhaps in the

adoption of printing typeface letters in an incorrect sequence in a variable printing stamp made for a single usage. A consolidated stamp that had been specifically made would be correct in its compounded setting and this makes one conclude that the defective Coade stamps must have been created individually using an arrangement of typeface single letters in a printing block almost as a one-off usage on each occasion.

- A further example of an very unusual and unique Coade stone impressed mark is provided by the Royal coat-of arms above the entrance doorway to 6, Suffolk Street, Westminster, the Royal Tailors to Queen Victoria: the mark is an impressed T. CROGGON upon the coat-of-arms of Queen Victoria, who did not accede to the throne until June 20th 1837. This Coade stone artefact must then post-date June 1837, and surely represents one of the last artefacts to emerge from the Lambeth manufactory as the lion statue made for the Lion Brewery is dated 24th May 1837! This item has been included in Table 1.3 but does not appear in Alison Kelly's survey (Kelly, *Mrs Coade's Stone*, 1990) and must be the only one that depicts Queen Victoria's coat-of-arms. However, a Coade stone female figure of an unknown provenance bears the stamp COADE LAMBETH 1840, which is extraordinarily late for the factory manufacturing operation, which by that time had formally closed down its production.
- Occasionally, Coade stone artefacts, such as statues and keystones, appear in auction sale catalogues and several of these do bear an undated mark, but these have not been included here as often their provenancing and source is not certain.
- The last Coade stone artefact with provenance manufactured with a dated and certified mark is the Lion Brewery statue dated 1837, although we have identified a Royal coat-of arms that must historically post-date this to later in 1837 or possibly early 1838 and of course we have noted the female garden statue bearing the date of 1840 but which comes without a provenance. The earliest Coade stone artefact made is rather more difficult to identify as the earliest impressed date mark is 1787, some 18 years after Eleanor Coade started producing Coade stone and there must surely be hundreds of exemplars that existed prior to this: however, it appears that one of her early successes was achieved through Robert Adam, who was engaged by the Duke of Northumberland to design an ornamental archway for his London home at Syon House, Islington. This was duly accomplished in 1770 but after a very hard winter the composite source material that had been used for the decorative statues and urns on the archway had seriously disintegrated and Adam needed to replace sections of the archway, which he did using the new Coade artificial stone in 1772. At that time Eleanor Coade is recorded as being incensed by the allegation that the deteriorated statues being replaced in this location were of her Coade stone, when they were actually made of an inferior composite material purchased from one of her competitors. Robert Adam nevertheless replaced the damaged artefacts with genuine Coade stone; these artefacts have survived to the present day at Syon House and act as testimony to the faith that Robert Adam had in Eleanor Coade's novel artificial stone material, which he went on to use elsewhere. The Syon House ornamental archway of 1772 must surely then be one of the earliest dateable examples of

Coade stone, even if there is no impressed date mark to confirm that assignment. It was an impressive construction and was replicated some years later in 1814 for the exiled Regent of Portugal at his Quinta da Boa Vista Palace in Rio de Janeiro, where it may be seen today spanning the entrance to the Jardim Zoologico. In this context, of course, the two Medici and Borghese Vases at Kedleston Hall, Derbyshire, which bear the undated stamps COADE LONDON and COADE LONDON Fecit can be reliably dated to 1771 from historical documentation records preserved in the house.

References

C.E.B. Brett, *Court Houses and Market Houses of the Province of Ulster* (1973), p. 74

E. Coade, A Descriptive Catalogue Of Coade's Artificial Stone Manufactory, At King's Arms Stairs, Narrow-Wall, Lambeth: Opposite White-hall Stairs. With Prices Affixed. Consisting Of, I. Statues. II. Busts. III. Vases. IV. Pedestals. V. Capitals. VI. Frizes. VII. Fascias. VIII. Panels and Tablets. IX. Medallions. X. Pateras for Elevations from 30 to 40 Feet. XI. Ditto from 20 to 30 Feet. XII. Pateras for Elevations from 10 to 20 Feet. XIII. Ditto of 10 Feet and under. XIV. Coats of Arms. XV. Imposts. XVI. Key-Stones. XVII. Architraves. XVIII. Rustics. XIX. Pinnacles. XX. Balusters. XXI. Balls and Feet. XXII. Chimney Tops. XXIII. Piers. XXIV. Consoles. XXV. Sepulchral Ornaments. XXVI. Chimney-Pieces. XXVII. Pieces of Furniture and various ornaments. XXVIII. Ornaments to fix on Stone or Wood. XXIX. Wood Chimney-Pieces with Stone Ornaments. With an Appendix, To be had at the Manufactory and of J. Strahan, Bookseller, No. 67, the Strand, London (1784)

K.A. Esdaile, Coade Stone. Archi. Build. News CLIX, 94–96 and 112–114 (1940)

K.A. Esdaile, *English Church Monuments, 1510–1840* (B.T. Batsford Ltd, London, 1946)

C. Fowler, Some remarks on terracotta and artificial stone as connected with architecture. Civil Eng. Archit. J. **13**, 215–216 (1850)

A. Kelly, *Mrs Coade's Stone* (Self-Publishing Association, Worcestershire, 1990)

M. Ponting, *Investigation of Coade Stone Fragments from West Dean College, Sussex*, Centre for Archaeology Report (Fort Cumberland, Eastney, Portsmouth) (English Heritage, 2003), pp. 1–5. ISSN: 1473-9224

C. Stanford, *Revisiting the Origins of Coade Stone* (The Georgian Journal, XXIV, 2016), pp. 95–116

The Gardens Trust, *Artificial Stone 2: Eleanor's Enterprise* (The Gardens Trust Publications, London, 2016)

H. van Lemmen, *Coade Stone* (Shire Publications Ltd, Princes Risborough, 2006)

Chapter 2
Factors That Influenced the Success of Coade Stone

Abstract The factors which influenced the success of Coade stone are considered; its resistance to environmental degradation, its versatility of artefact manufacture, the multiplicity of provision of statuary and of building items, and especially the ease with which the production of heraldic emblems such as coats-of-arms and escutcheons to specific sizes are noteworthy. Some specific examples are used from the Coade & Sealy Catalogue of 1784 to illustrate. Restoration problems of badly damaged Coade stone artefacts using modern materials are discussed. The demise of Coade & Co. is attributed to the appearance of cementitious artificial stone alternatives and the later manufacture versions by John Blashfield, Mark Blanchard, Austin & Seeley, James Parker and James Pulham. Confusion of terminology with "terracotta" artefacts in the mid- to late-nineteenth century.

Keywords Heraldic emblems in Coade stone · Coade & Sealy catalogue · Portland cement · Blashford · Blanchard · Austin & Seeley · Pulham rock · Terracotta

It is appropriate here to consider what factors influenced the notable success of the Coade artificial stone business within a year or less of Eleanor Coade taking over the ailing business of Daniel Pincot in 1769. Firstly, the manufactory was not enlarged or moved and a warehouse advertising their wares with their Catalogue does not appear until many years afterwards—in 1784. Secondly, several historians have alleged that Eleanor Coade must have refined and improved Daniel Pincot's recipe and formulation for artificial stone and that this was the major reason for the success of her business when she had purchased the ailing manufactory from Daniel Pincot in 1769 and turned its fortunes around so rapidly. However, background research reveals that Eleanor Coade would not have had the required scientific background or experience in ceramics production to enable her to achieve a significant improvement in stone composition even from an established recipe over such a relatively short time. In an analogous scenario, the early porcelain manufacturers did experiment with their recipes, formulations and processing conditions and it is clear from the records that this involved much sustained effort, with a detailed and organised stepwise approach to recipe refinement, process control and the services of people who were scientifically competent in managing the changes and interpreting the

H. G. M. Edwards, C. J. Brooke, *Coade Stone*, Cultural Heritage Science,
https://doi.org/10.1007/978-3-031-71432-0_2

results of the largely empirical experiments that were being undertaken to effect an improvement in the ceramic product (Edwards, *Porcelain Analysis and Its Role in the Forensic Attribution of Ceramic Specimens,* 2021). For example, Lewis Weston Dillwyn, a Fellow of the Royal Society and the owner of the Swansea China Works, undertook a series of recipe changes to his porcelain body composition from 1814 to 1817, aided by Samuel Walker, who was regarded as the finest kiln manager in the country at that time and who had been engaged by Martin Barr of the Worcester China Works to undertake some advanced experiments in reverberatory kiln design. Dillwyn was also an eminent and well-respected scientist, elected an FRS in 1804, and he therefore had access to some of the finest brains and practical experience in the ceramics industry (for example, Sir Henry de la Beche, Sir Joseph Banks, William Billingsley, William Weston Young and others)—yet his careful experiments still took almost two years to undertake and evaluate to produce a fine porcelain body (Dillwyn, *Notes of Recipes for Porcelain Bodies and Glazes 1815–1817,* reproduced in Eccles & Rackham, *Analysed Specimens of English Porcelain in the Victoria and Albert Museum Collection*, 1922). He did achieve this eventually but it certainly was not a simple passage and there were several setbacks along the way. In contrast, Eleanor Coade did not seem to have any formal scientific training at all, although it seems that Charles Fowler, a distant cousin, did imply this without giving any evidence for its existence (Fowler 1850), perhaps trying to explain her remarkable success in turning around a failing manufactory? What is definitively clear and unambiguous is that Eleanor Coade was an astute business woman and she appreciated very well her clientele requirements, which she was able to deliver by employing the finest designers, sculptors and architects of her day (Stanford 2016). She must have acquired some keen business knowledge and acumen from her previous foray into the linen textile business and also had thereby amassed the requisite finance to buy Daniel Pincot's failing business at the appropriate time in 1769. It can be no coincidence that at this time the landowners and aristocracy, many of whom had undertaken the Grand Tour of Europe, were also keen to acquire embellishments and statuary for their houses in the neo-classical style of the Greeks, Romans and Egyptians for which they engaged architects to design and rebuild their homes and estates. This coincided with the engagement by Eleanor Coade of architects of the calibre of Robert Adam, John Nash, Sir John Soane and Lancelot "Capability" Brown, all of whom were enthusiastic to use Coade stone in their projects with its wonderful array of standard and bespoke neo-classical statuary and artefacts. The major advantage for this purpose of a moulded fired ceramic artificial stone for pediments, pilasters, columns, friezes and paterae was that it could be more readily produced in quantity and was much cheaper than the natural sculpted stone. This was immediately obvious and the growth in the desire for neo-classical garden statuary provided in the same medium for a wide range of clients was apparent. Even later, in 1815, William Croggon, working then with Eleanor Coade, was engaged by the Imperial War Museum, through the architect Humphrey Repton, to manufacture a large Royal coat-of-arms some 15 feet in length, for which he quoted the price of 130 guineas (£136-10s-0d), which was significantly cheaper than the quotation they received at that time for a sculpted version in Portland stone costing £500. The Imperial War Museum, therefore, had no qualms about engaging Eleanor Coade and Coade & Co. to produce this item in Coade stone.

The claim that the artificial Coade stone was also more resilient to environmental erosion and general weathering than its natural congeners was also seen to be true and today, some two hundred and fifty years later, the demonstrably better preservation generally of the Coade stone artefacts that were mounted upon natural stone substrates is often visibly significant as in most cases the natural stone has suffered badly in comparison, becoming pitted and subject to surface spallation from the seasonal temperature melt/thaw cycles that are so environmentally damaging to naturally porous materials (Scott, *Artificial Stone: A Successful Substitute for Natural Stone?*, 2009). Of course, the ceramic artificial stone artefacts do become stained with the metal salts of iron and copper artefacts that were mounted in close association with them, with salts deposited from aerial and wind-borne pollution and are subject to the biological invasion of lichens and cyanobacteria, as does natural stone, but these blemishes are strictly locally confined to the surface and can be removed with assiduous cleaning, whereupon the underlying surface of the ceramic stone is seen to be generally well preserved. In a related study of the lichen invasion of *Dirina massiliensis* forma *sorediata* of marble surfaces in the Palazzo Farnese in Caprarola, to the north-east of Rome, which was built by Vasari in 1560, the hyphae of the lichen colonies have penetrated over 15 mm into the calcite marble substrate and the oxalic acid produced in their metabolic cycle has thereby converted approximately one kg of calcite into calcium oxalate monohydrate and dihydrate, $CaC_2O_4.H_2O$ and $CaC_2O_4.2H_2O$, for every square metre of surface that has been exposed to the organism (Edwards et al. 1991, 1992; Edwards and Seaward 1993). This has resulted in a complete destruction of surface sculpture and motifs in relief, as well as the applied pigmented frescoes, which would not have happened with the Coade stone analogue substrates. It should not be assumed, however, that all Coade stone artefacts were uniformly standard in this respect as information is now appearing from restorers that sometimes their quality appeared to be somewhat variable, which might be a reflection of the operating thermal conditions and the actual placement of the items in the kiln, as does their analogous porcelain artefacts (Philip Thomason, Cudworth, personal communication).

The stabilisation and restoration of Coade stone artefacts has generated some divisive issues regarding authenticity which have been reviewed recently by Rix and Emmitt (2023) in a paper which highlights the unique challenges represented in the conservation of artificial materials compared with their natural analogues, wherein scagliola, Coade stone, marbles, marezzo (a later version of scagliola using dyed silk threads to simulate veining in in a gypsum plaster) and graining were specifically addressed. The use of modern substances that are considered equivalent to those used in the original artificial manufacturing versus those that are more compatible with the original materials themselves needs a strategy for the establishment of authentic conservation protocols. The restoration of the King George III statue on the promenade at Weymouth, which we shall consider in more detail as a case study in Chap. 7, is cited as an example of inappropriate intervention which has resulted in its current appearance being totally unlike that which was envisaged originally for the statue in 1810: originally bronzed, the Coade stone statue was restored in 1948 and painted in polychrome pigments and this was not as it was designed, and has since been further conserved in a more recent restoration programme. Undoubtedly, this aspect of conservation of ancient artificial materials versus perceived unsympathetic modern imitative approaches will be the subject of more discussion in the future around Coade stone and related materials.

The scholastic work by Alison Kelly on Coade stone (Kelly, *Mrs Coade's Stone*, 1990), in which she has identified over 700 existing artefacts and locations where these could be viewed in 1990 also gives a list of 148 architects who used this material in their projects between 1769 and 1837. Robert Adam was certainly one of the first to adopt this new material in preference to natural stone in the early 1770s when he redesigned the ceremonial gateway to Syon House in Isleworth, the London home of the Duke of Northumberland, which had experienced some environmental degradation since it was first erected: today, the Coade stone Northumberland lion atop the gateway and the associated plaques and paterae can be seen to have fared much better over the years than have their natural Portland stone substrates. Included in the list of architects who favoured the use of Coade stone in their projects are: Robert Adam, Sir Charles Barry, Lancelot Brown, George Byfield, Sir William Chambers, John Johnson, Thomas Leverton, John Nash, Giovanni Piranesi, Humphrey Repton, Sir John Soane, Charles Tatham, Sir Robert Taylor, James Wyatt and Samuel Wyatt (Kelly, *Mrs Coade's Stone*, 1990). In particular, Robert Adam, one of the first architects to use Coade stone in his projects, used plaster for his designs of features and embellishments inside houses but invariably chose Coade stone for his external designs, which comprised delicate bas-reliefs, grotesque decoration, neo-classical figures, friezes, swags and paterae in addition to garden statuary (Bolton, *The Architecture of Robert and James Adam*, 1922).

Perhaps the most significant initial business move made by Eleanor Coade was her engagement of distinguished sculptors and modellers to produce the original works from which the moulds could be made for the initial manufacture of the artificial stone artefacts: these included Benjamin West, Joseph Panzetta and John Bacon. Indeed, so successful was Eleanor Coade in her acquisition of their exclusive sculpting services to her business that Josiah Wedgwood commented in 1780 that he was unable to break into the burgeoning novel business of ceramic artificial stone manufacture at that time because Eleanor Coade had already secured the services of the major designers, sculptors and architects: this was praise indeed and came from perhaps one of the most astute, experienced and successful ceramics manufacturers in England who was regarded by his competitors as one of the most influential ceramicists in his field since he set up his family ceramics business in Etruria, Staffordshire, in 1759.

Coade stone was a particularly popular material for the construction of heraldic devices especially crests and coats-of-arms, for great houses as well as funerary monuments (K. Esdaile 1940a, b) as it was possible to execute complex patterns with a stylus into the still moist paste of the moulded artefact after its removal from its mould and prior to its undergoing its final firing sequence in the kiln. Intricate patterns could be easily and accurately carved into the stone paste using a stylus, which upon firing would then be permanently registered into the artefact: in this way it was possible to introduce the detailed geometric designs required for the accurate depiction of the heraldic features on the escutcheon of a coat-of-arms which was necessary to reflect the heraldic tinctures involved (a process known as "tricking") that were so critical to the proper representation of the coat-of-arms of the arms-bearing family. These were defined using a series of lines, dots and hatchments to indicate the main heraldic tincture colours of red (gules), blue (azure), green (vert), yellow (or), white (argentum), black (sable) and purple (purpre) that were borne on the polychromatic heraldic shield and were also used in the engraving of the escutcheon or coat-of-arms on silver plate belonging to an armigerous individual (Edwards, *Welsh Armorial Porcelains: Nantgarw and Swansea Crested China*, 2022). In the sculpting of coats-of-arms in marble and natural stone this was an extremely difficult task to accomplish and it could take many hours of painstaking work to achieve the result. A very good example of such a coat-of-arms modelled in Coade stone that demonstrates this intricacy of detail in the modelling is that of General Lord Robert Manners, a grandson of the Duke and Duchess of Rutland, who commissioned his heraldic bearing in Coade stone to be set above the gateway to his estate at Bloxholm Hall in 1772 (Fig. 2.1). His crest was a peacock, which was also modelled separately in Coade stone in 1772, and assumed pride of position above the

Fig. 2.1 View of Bloxholm Hall from a print of 1825: the gable to the main house bears a coat-of-arms with the peacock crest and unicorn supporters in Coade stone. The crest and other associated heraldic devices did not survive the demolition of the house in 1963

Figs. 2.2 The coat-of-arms of General Lord Robert Manners, who inherited the estate from the Duchess of Rutland in 1772. He commissioned the architect, Lewis Vulliamy, to redesign the house and incorporate Coade stone features. This coat-of-arms was removed from the gateway of Bloxholm Hall in 1812 when General Manners remodelled the church of St Mary's at Bloxholm and was installed above the entrance to the vestry

central gable to the house itself along with another coat-of-arms and supporters. The main house was demolished in 1963 but General Lord Manners' coat-of-arms had already been transferred to St Mary's Church, Bloxholm, in 1818 where it now sits above the vestry entrance (Fig. 2.2). The peacock crest of his coat-of-arms can also be seen here atop the escutcheon but unfortunately the reputedly very fine free-standing Coade stone peacock crest which had also been commissioned and which was admired by visitors to Bloxholm Hall for many years has now disappeared and was presumed destroyed during the demolition process of the house that was carried out some years later. The horizontal hatchings drawn on the two bars of the escutcheon on a dotted hatching background indicate that the bars were heraldically azure (blue) and the background or (yellow). On the bottom left of the arms is impressed COADE & SEALY, LAMBETH, 1772. The crest in Fairbairn (Fairbairn, *Crests of the Families of Great Britain and Ireland and Their Kindred in Other Lands,* Volumes I and II, 1905) for Robert Manners is confirmed as a peacock in its pride ppr. (shown in Plate 103/5 therein) and the escutcheon is shown in Burke's Peerage (Sir J. Bernard Burke, *Burke's General Armory—The General Armory of England, Scotland, Ireland and Wales: Comprising a Registry of Armorial Bearings from the Earliest to the Present Time*, 1884) where it is described as: Or, two bars azure, a chief quarterly of the last and gules, 1st and 4th two fleur-de-lis, 2nd and 3rd a lion passant gardant, or. The supporters are two unicorns, maned and futted, unguled, or. This certainly matches very accurately with the Coade stone artefact as shown here but interestingly it seems that the two unicorn supporters have suffered some physical damage over the years, perhaps during the relocation of the coat-of-arms to Bloxholm Church and have now lost their horns and a foreleg! The Manners motto is also shown beneath the escutcheon as *Pour y Parvenir*, which translates as *Strive to Attain.*

Whilst on the subject of heraldry and Coade stone, it should be mentioned that the gateway to Belmont House in Lyme Regis, which was bequeathed to Eleanor from her uncle Samuel Coade in 1784 and which she used as a holiday villa, thereafter referring to it as her "seaside pavilion", displays a crest with a coot in Coade stone, which has been assumed by local historians to be a rebus or a pun on the name Coade (see Fig. AI.2) (Hansford 1960–1961). It appears, according to George Roberts (*The History and Antiquities of the Borough of Lyme Regis and Charmouth*, 1834, p. 296), that John Coade, a wealthy relative of Eleanor's and mayor of Lyme Regis, may have adopted this device as a pun on the name of Coade, which has several spelling interpretations in local dialect, including Cood (which was pronounced locally as Coot): it is unclear, however, when this crest was commissioned and by whom because historians maintain that Eleanor herself made the Coade stone additions to Belmont House upon her inheritance. Whatever the chronology, John Coade did not have the right to bear arms and no armigerous individuals with the related family name Coade appear in Fairbairn: local historians claim that some West Country families with the name Coade adopted different crests including a Cornish chough, but again there are no spelling variants of Coade, Cood, Coot or Coote that have armigerous bearings described in Fairbairn (Fairbairn, *Crests of the Families of Great Britain and Ireland and Their Kindred in Other Lands*, Volumes I and II, 1905). It can be concluded therefore that the Coade family individuals did not have the right to bear arms and that the Coade stone device above the gateway to Belmont House is bogus, although it is listed in Alison Kelly's work (Kelly, *Mrs Coade's Stone*, 1990) as the family crest (which apparently it seems they were not entitled to use!).

Eleanor Coade was very proud of her neo-classical repertoire of artefacts, which could be "tuned" to the personal wishes of individual clients with the addition of motifs: in the existing records, occasionally one meets with the specific need to make an individual sculpture in soft clay from which a mould would then need to be made for making the Coade stone replica. This naturally incurred an additional cost to the client initially, which in some cases matched the cost of the Coade stone replica itself, effectively thereby doubling the overall expense of the commission. It is interesting too that where invoices exist in family records, the cost of provision of the Coade stone artefact more often than not exceeds the stated cost quoted in the Coade & Sealy Catalogue of 1784 and published as E. Coade and J. Sealy, *A Descriptive Catalogue of Coade's Artificial Stone Manufactory at King's Arms Stairs, Narrow Wall, Lambeth: Opposite Whitehall Stairs. With Prices Affixed,* which was available at the Manufactory and at J. Strahan, Bookseller, 67, The Strand, London, 1784 (see Appendix II). A book of etchings of Coade designs is now archived in the British Library, *Etchings of Coade's Artificial Stone Manufactory at King's Arms Stairs, Narrow Wall, Lambeth,* 1784, British Library. The Coade stone paste statuary was constructed in the moulds in parts and then assembled whilst still soft, using interior armatures to provide strength if needed, and the complete statue was then fired as one piece in the kiln.

As an example of differential pricing, the sundial provided for Weston Park in Shropshire in 1779 is invoiced at £28 in the house records, which far exceeds the listed cost of 18 guineas in the Coade Catalogue, namely £18-18s-0d. It is not clear wherein lies the origin of this discrepancy in the pricing: it could be that certain additional motifs were requested by the client, and that there was perhaps an additional cost for the brass gnomon and its base (although the invoice refers to the sundial plinth itself) or perhaps there was an additional transportation charge levied for the journey from London? On the other hand, the Medici Vase ordered directly from Eleanor Coade by the Earl of Carlisle for Castle Howard in Yorkshire in 1778 was charged at £18-0-0 (and not guineas!) in the family records against the list price of £31-0-0 which appeared later in the Catalogue, which is inexplicable unless one invokes this as perhaps being a trial early artefact prior to the establishment of the first Catalogue and associated list prices which only appeared in print some six years later in 1784? It has been noted that William Croggon's invoices for Coade stone provision in his commissions sometimes mention that carriage is included, with the statement "*waggons provided*", so clearly this could be a significant contributing factor to the overall price when houses or delivery sites situated some distance from London were involved for the transportation of the artefacts. Perhaps the most expensive carriage fees recorded for a commission in Coade stone were incurred by the Duke of Northumberland in 1812, who ordered several statues and an ornamental gateway which was a copy of his own Brentford Gate at Syon House but embellished with the Portuguese Royal coat-of-arms, to be installed at the Quinta da Boa Vista Palace, Rio de Janeiro, for his friend the Regent of Portugal, who had escaped Napoleon's invasion of Portugal and was at that time exiled in Brazil and who wished to have an example in his estate there. Eventually, the Duke had to settle the amount of the invoice personally for these Coade stone items, their carriage to Brazil and for the services of a stone mason to travel and erect them in Brazil: an episode which Alison Kelly has recounted in her article entitled "An Expensive Present" (Kelly, *The Burlington Magazine*, 1984)!

Eleanor Coade's pride in her classical statuary is evidenced by the portrayal of a classical scene at the entrance to her own Gallery at Westminster Bridge in 1799 as drawn by John Bacon to advertise to potential clients the location of the Gallery (Fig. 2.3) which shows:

> Fire, with flourishing hair and a flaming torch in her hand leaping aggressively at a recoiling Father Time and gripping his arm, causing him to lose the hold of his scythe.

The Latin phrase from Ovid above her kiln entrance describes the Preservation of Antiquity through Fire: *Nec edax abolere vetustas* ... meaning, *Devouring but not destroying antiquity*. A print of the Coade kiln interior with the legend "*View of the River God THAMES, 9 Feet figure, & 4 Seasons as placed in the kiln, for Burning so as to represent stone, at the Lithodipyra at Lambeth*" can be seen here (Fig. 2.4) as drawn for the *European Magazine* in 1786 by J. Sewell, Cornhill, London.

Fig. 2.3 Drawing for a business card made by John Bacon to advertise the location and contents of Eleanor Coade's Gallery in 1799 to potential clients. Note the Latin inscription above the entrance to the kiln: *Nec edax abolere vetustas*, meaning *Devouring but not destroying Antiquity*. This is missing from the kiln diagram shown in the next Figure

2.1 The Demise of Coade Stone

Irrespective of the development of the alternative earlier precursors of high-temperature fired Coade stone, it is relevant now to evaluate the artificial stone composites that appeared contemporary with and after Eleanor Coade had established her manufactory. In January 1767, Daniel Pincot is recorded as opening his manufactory first at Goulston Square, Whitechapel, then this was taken over in the autumn of the same year by George Davy, Pincot meanwhile had moved his operations across the river to the King's Arms Stairs, Narrow Wall, Lambeth. It seems that Pincot did not have a showroom for his wares but he sold his artefacts that were "*made in several Compositions*" directly through auction rooms in Pall Mall and he set up his home at 18, The Corner, Glastonbury Court, Long Acre. He was certainly not alone in the artificial stone business as a Samuel Arnold is also recorded as being still in business in Golden Square, London in 1780 and a William Brooks in Lambeth also was manufacturing artificial stone at this time (see later), as well as George Davy. Pincot claimed that the demise of Richard Holt's business in the early 1730s, the date of 1732 is usually given for this, arose from "*the Projector dying,*

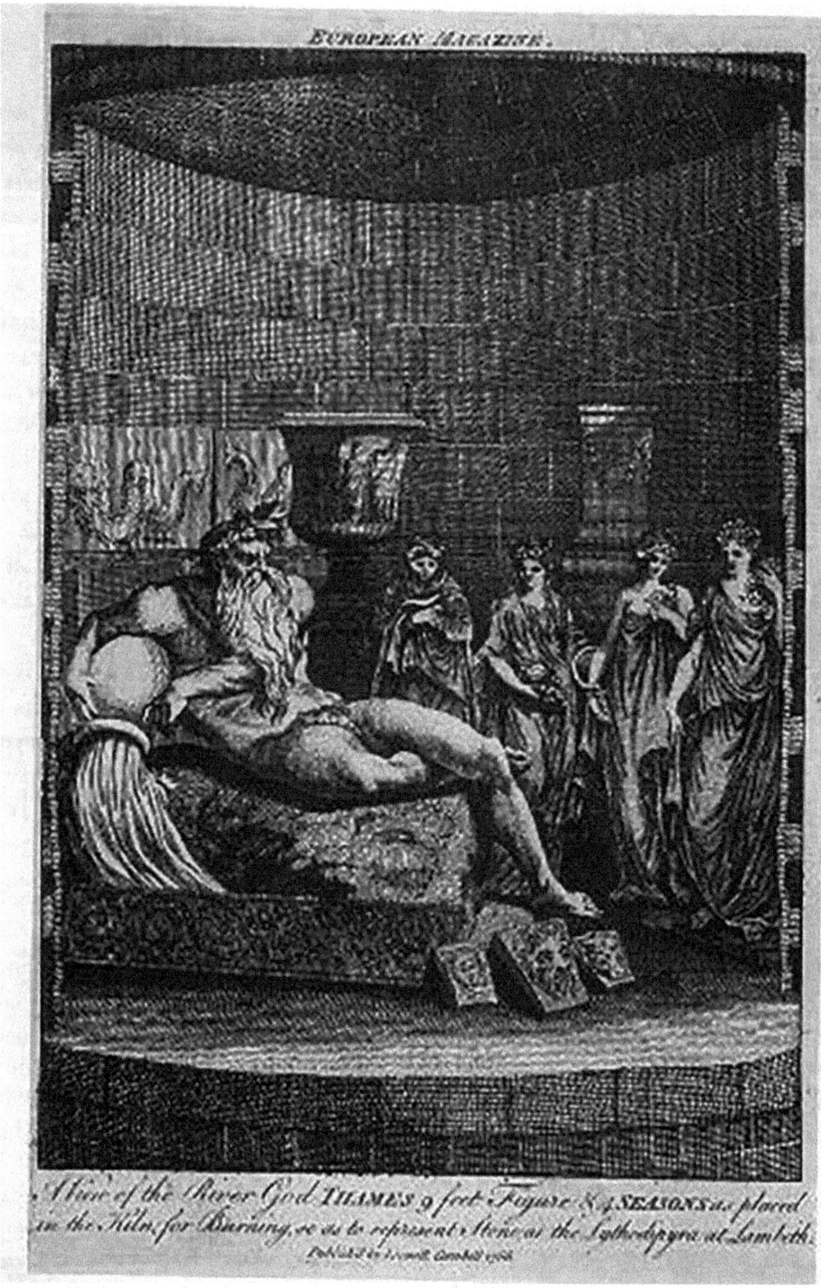

Fig. 2.4 Print from the *European Magazine* of 1786, showing the interior of the Coade Kiln at Lambeth, with the legend: View of the River God THAMES, 9 Feet Figure, & 4 Seasons as placed in the kiln, for Burning so as to represent stone, at the Lithodipyra at Lambeth. (Published by J. Sewell, Cornhill, London)

when the whole Affair died also". In 1770, Daniel Pincot wrote a treatise on his artificial stone (Pincot, *An Essay on the Origin, Nature, Uses and Properties of Artificial Stone*, 1770) in which he claimed unsustainably that its synthesis had been accomplished in prehistoric times and had been used by the ancient Egyptians to build the pyramids and by the Chinese to build the Great Wall.

Much of the earlier business activities of the artificial stone business in London was detailed in the *Daily Advertiser,* for which a complete copy run exists in the Library of Congress, Washington DC, USA, which had been purchased by subscription by Thomas Jefferson (1743–1826, Secretary of State under George Washington and who became the 3rd President of the USA between 1801 and 1809) and was used there to ascertain specifically what items were then being made industrially in London. Nancy Valpy has comprehensively surveyed the news in the *Daily Advertiser* over the period from about 1745 until 1780, about who was trading and manufacturing in London, and her conclusions were initially presented in two papers to the Linnaean Society, Burlington House, London on the 12th January and the 16th February 1985. These were later extended back to 1731 with input from Alison Kelly (Valpy and Kelly 1986). This survey is a mine of information and gives details of who was manufacturing artificial stone in London at that time. Prior to her taking over Daniel Pincot's business in 1769, Eleanor Coade had earlier acquired artificial stone manufacturing premises in Knightsbridge which she later disposed of in 1778 in an announcement in the newspaper on the 24th December, when her acquisition of Pincot's business in King's Arms Stairs, Lambeth, had been well established for several years. On the 11th September 1771, the presence of Daniel Pincot in Eleanor Coade's business was terminated as she had learned of his false claim publicly that he was promoting himself as the motivating partner in her business, announcing in the *Daily Advertiser*, that "*Mr Pincot has no Proprietary in this Affair*". This was reiterated in a further announcement on the 14th September, 1771. On October 23rd 1771, Eleanor appointed the noted sculptor John Bacon as her factory manager, whom she termed a Superintendent, and her artificial stone business was then truly launched. She was able to announce in March 1773 that her successful business was "*Employed by many of the principal Nobility and finest Architects in the Kingdom*". Despite many historians recording that John Sealy was appointed partner of Coade & Co. in 1771/2, there is no mention of this at all in the *Daily Advertiser*, or of his involvement in future years of her business. However, on the 24th December 1778, it is recorded that Eleanor Coade had closed her artificial stone factory at Knightsbridge and was then concentrating her business at King's Arms Stairs, Lambeth. With the termination of George Davy's business in the early 1770s the only other artificial stone manufactories recorded in London were then those of Arnold & Co. of No.1 Upper James Street, Golden Square, recently split from John Lynn and Thomas Sharpe, who had set up in their own business which was still operating in an entry made for 23rd February 1780, and William Brooks at Prince's Street, Lambeth. Clearly, none of these contemporary competitors of Eleanor Coade in artificial stone manufacture was able to compete successfully with her product, yet they were still operating for over ten years after she had taken over Daniel Pincot's business. It is quite possible that they were offering an inferior product to Coade stone more cheaply that was still useful economically and aesthetically for some purposes to prospective clients.

Daniel Pincot has claimed that he did not adopt Richard Holt's formulation and recipe for artificial stone and this statement needs to be evaluated analytically. What is interesting from the auction sales announcements in the *Daily Advertiser* is that both George Davy and Daniel Pincot were offering artificial stone artefacts in a range of composites, which implies that they must have had several different recipes for the production of artificial stone at their manufactories.

2.1.1 Roman Cement and Cementitious Stone Manufacture

In the 1780s the Rev. James Parker started developing a composite cement which he patented under the name of "*Roman cement*" in 1796 and formed a business venture (called Parker & Co. in the same year) using this composite for the "*Manufacture of Coping Stones, Window Sills, Balusters, Blocking and String Courses, Gothick Ornaments and Others of this Nature Cheaper than can be executed in Natural Stone*". The appellation of "Roman cement" to Parker's synthetic material is rather misleading as this historically had been made by the Romans by burning *septaria,* which comprised natural nodules of marl (also called "cement stone", an argillaceous marble) found in clays which contained clay and calcite in admixture. This was burnt at a temperature of around 600 °C then ground and mixed with salts to produce a quick-setting mortar, which cured over some 5–15 min after the mixing. Various types of cement were in use in Roman times, which were given different names according to their formulations: for example, *opus caementicum* comprised crushed rock and burnt lime as a binder for volcanic ash (*pozzolan*) and pulverised bricks which was then used as *cementium,* their fabled hydraulic cement used for the waterproof lining of their aqueducts and water storage areas. The Rev. Parker's patent certainly was nothing like this in composition yet he pronounced that it was "*a certain cement or terracotta to be used in aquatic and other building and stucco work*". Essentially, it seemed to be a mixture obtained by burning powdered bricks, tiles and chalk, the elevated temperature converting the chalk to calcium oxide or quicklime, which provide the flux and binder for the final composite aggregate when mixed with water. This was not a very successful alternative to Coade stone as it was predictably of a rather unattractive brown colour and was judged at the time to be suitable only for use in "hidden locations". The major step forward in cementitious artificial stone production came when Joseph Aspdin (1778–1855), a Leeds brickmaker, took limestone and heated it in a kiln to temperatures of near 900 °C and formed a clinker aggregate which was then ground with the addition of 3% gypsum, calcium sulfate dihydrate. He patented this process on the 21st October 1824 and called his material "*Portland cement*" because of its resemblance on setting to Portland limestone which was quarried on the Isle of Portland in Dorset and which was finding much favour with many architects for important and prestigious buildings of the time. Apparently, he had actually discovered the process for making his cement some years earlier, perhaps even as early as 1799, but needed to perfect the manufacture and set up his business. Joseph Aspdin first sold his new composite Portland cement from premises in the Angel Inn Yard in Leeds and named it after the natural stone that it was intended it should imitate. However, the new composite did not appeal widely to prospective users until his son William Aspdin refined the mixture, developed it over the succeeding years, and made a concrete composite using the basis of Portland cement mixed with a hydraulic lime, gravel and sand which only required the addition of water at ambient temperature to achieve a set (Francis, *The Cement Industry, 1796–1914: A History*, 1977). Unlike Coade stone, therefore, this cementitious artificial stone was cold-cured and not high-fired and therefore was not a ceramic composite—the saving in fuel costs and kiln operations over the prolonged heating and cooling cycles that were necessarily demanded for the Coade stone artefacts was nevertheless significant and contributed to the cementitious stone appeal as a more economical option. Also, the setting process at ambient temperature of the Portland cement composite meant that the moulds could be used on site and this was more economical for transportation and for raw materials availability, and was especially favourable for building work on site involving the manufacture of cement blocks and ashlars. Although the Romans perfected the use of cement in their construction industry it seems that they never actually used it with moulds to produce artefacts and building blocks and the first recorded use of a true Roman cement in this context was found in the ancient mediaeval walled city of Carcassonne, where cast cement lintels of a pozzolanic Roman cement have been found dating from the mid-twelfth century. The earliest recorded use of an artificial stone composite in Britain can be ascribed to King Henry VIII (1491–1547), who employed his brewer to use artificial stone composites in the reconstruction of stone mouldings and bricks for his royal palace at Sutton Place.

This early Portland cement of Joseph Aspdin should not be confused with the modern variety of the same name, which is fired at a higher temperature and which has the general chemical composition as follows: $3CaO.SiO_2$ (C3S, alite) 55%, $2CaO.SiO_2$ (C2S, belite) 20%, $3CaO.Al_2O_3$ (C3A, celite) 10%, $4CaO.\ Al_2O_3.\ Fe_2O_3$ (C4AF, brownmillerite) 10%, and $CaSO_4.2H_2O$ (CS2, gypsum) 5%. This cement is still, of course, mixed with sand and gravel or pebbles and water to make concrete: typical analytical data for Portland cement can be measured as follows—SiO_2 22%, Al_2O_3 7%, Fe_2O_3 3%, CaO 63%, MgO 3% and SO_3 2%. These data will be relevant when we come to consider the analytical data for Coade stone artefacts that have been restored in modern times using composite mixtures featuring Portland cement that can be differentiated from their relatively very high calcia content, in complete contrast with the calcia content of Coade stone, which is estimated to be around 1% or less. Modern "replacement" composites for Coade stone used in restoration projects vary in composition but a favoured one with restorers seems to be Haddonstone (TecStone), a semi-dry cast limestone composite with a surface texture similar to Portland stone and which comprises 50% crushed limestone with a particle size of around 1–10 mm mixed with finely crushed limestone aggregate, cement as a binder and waterproofing and appropriate pigment additives to simulate Coade stone. Other artificial stone materials from the nineteenth century that require restoration include *scagliola*, and also pigmented stucco and artificial lime-based marbles such as *marezzo* and *marmerino* (Rix and Emmitt 2023). Clearly, the inclusion of cement as a component will affect the higher elemental percentages detected of Ca and S, especially arising from its gypsum content. The perceived unsympathetic restoration of artificial materials using modern resins and equivalents has been mentioned earlier in connection with the painting of a Coade stone statue of King George III which was never intended to have been so treated originally.

By the 1840s, the Aspdin Portland cement composite had achieved a much wider appeal and was then starting to be adopted in earnest for building projects, which correlates chronologically with the demise of William Croggon's Coade stone business failure in the mid to late 1830s. After William Croggon died in 1835 he was succeeded in the artificial stone business by his son, Thomas John Croggon, who managed to keep the Coade manufactory going for several years in the face of stiff competition from the cementitious stone variants. During this period, James Goding established his Lion Brewery on a neighbouring plot of land in Lambeth near the Hungerford Bridge leased by the Archbishop of Canterbury. Goding commissioned three Coade stone lions for his brewery from Thomas Croggon in 1837, the largest being some 12 feet tall and weighing 14 tons. The Lion Brewery closed in 1924 and the site was then absorbed as a storage facility by Hoare & Co of East Smithfield: the buildings were seriously damaged in a fire in 1931 before becoming derelict and then being demolished finally in 1949. The site was selected for the Festival of Britain in 1951 and during the demolition process the largest Coade lion was adopted by British Rail for the concourse of Waterloo station and painted in its corporate red colour. The personal intervention of King George VI helped save the large lion upon its removal from Waterloo concourse and a plaque on its plinth on Westminster Bridge now records that fact. A second, smaller lion was also saved and donated to the English Rugby Football Union headquarters at Twickenham in 1971, where it was gilded in 1991 for the hosting of the Rugby World Cup. It is said that the Twickenham lion actually represents the British and Irish Lions rugby team, as the emblem of the English Rugby Union Football team is the red rose. The lion still remains there and an interesting statistic is that prior to its installation at the Twickenham ground the England RUFC won 46% of their matches there but afterwards this figure rises to 73%, so the magic associated with the Coade stone lions is still apparently manifest! The cost of gilding the Twickenham lion in 1991 was £6000, which far exceeded the price paid for the Twickenham stadium site originally in 1901. The third lion from the Lion Brewery in Lambeth which sat above a third gateway has since disappeared and it is suspected that it was destroyed in the demolition process of the brewery. In 1966, the development of Waterloo station necessitated a further relocation of the large lion from the Waterloo station concourse and it was moved to its current site on a granite plinth at the eastern end of Westminster Bridge. During the move a small recess was discovered in the back of the lion which held a bottle containing two coins of King William IV and a trade card advertising Coade stone. In replacing this time capsule, further additions of some coins from 1966 and some documents were sealed into the cavity, one of which was an account of Coade stone written by J.H. Holdroyd for an issue of *The Times* on 5th

March 1966. During the earlier move of the lion statue from its Lion Brewery site in 1949, an impression was found underneath one of the paws of the lion confirming that it was sculpted by William Frederick Woodington and dated the 24th May 1837 and mentioned *Coades* in script. It must surely be recognised that the *South Bank Lion*, as it is now known, must represent one of the last pieces of Coade stone to have been manufactured at the Lambeth works and it is interesting that there does not appear to be an impressed mark attributing it to Coade & Sealy as had still been used some years earlier by William Croggon for some pieces of his own manufacture. Further details relating to these Coade lions and their location will be considered in a Case Study later.

As mentioned earlier, William Croggon purchased the Coade artificial stone business in 1821 upon the death of Eleanor Code and he continued to run it as a successful business, employing the same designs, moulds and architects for the installation of the artefacts. An advertisement in the *Daily Advertiser* of the 27th December 1821 announced that William Croggon & Co., as they were now known, still occupied "Coade's Ornamental Stone and Scagliola Marble Works at Narrow Wall, Lambeth" and that he had taken into partnership John Danford, a stonemason and statuary designer at Palace Row, New Road, and Belvedere Road, Lambeth. On the 2nd October 1832 in the *London Gazette*, Croggon announced that the partnership with Danford had been dissolved on the 27th August of that year. Shortly afterwards, William Croggon became bankrupt in 1833 (*Chelmsford Chronicle*, Friday, 11th October 1833) and the sale of his business interest then commenced in 1835 (*Morning Advertiser*, Tuesday 39th June, 1835; Roberts & Godfrey, *Survey of London: Lambeth, South Bank and Vauxhall*, **23**, pp. 58–61, 1951). It seems that Thomas Croggon, William's son, did try to recover the business from bankruptcy but he did not succeed and the Belvedere Road premises were let in 1837 to Thomas Routledge and John Danford (!) Greenwood, who as Routledge & Greenwood made terracotta and scagliola artefacts there but definitively not any items specifically of Coade stone. We may well pose the conjectural question now as to the logistics of the manufacture of the Lion Brewery lions (one of which, the South Bank Lion, carries the date of 24th May 1837 upon it). Coupled with this is the Royal coat-of-arms of Queen Victoria in Coade stone above the Queen's tailors at 6, Suffolk Street, Westminster, with an impressed mark of T.CROGGON, LAMBETH, which can be dated to after the 20th June 1937, when Victoria, at the age of 18 years and barely one month ascended to the throne upon the death of her uncle King William IV—her coronation would not be held until the 28th June 1838. It seems then that Thomas Croggon was probably responsible for creating the definitive final pieces of Coade stone to emerge from the works in Lambeth in 1837—and it is doubly interesting that one of the most cited artefacts in Coade stone, namely, the *South Bank Lion*, is one that must have originated from the Lambeth factory during his tenure there.

Certainly, by the early twentieth century, the appreciation of the superior environmental resilience of Coade stone that was offered in comparison with its natural stone congeners was realised and prompted the following statement by Yorke in the

Architects Journal (1932) which was echoed by Simon Scott (Scott, *Artificial Stone: A Successful Substitute for Natural Stone*, 2009):

> Comparing Bath stone to its artificial equivalent identified that the reconstructed material is much more durable and has not natural bed so that there can be no flaking of the surface. It is unaffected by sea air and the face is not destroyed by a sulphurous atmosphere.

2.2 The Cementitous Artificial Stone Manufacturers

We have already referred to the Rev. James Parker and his ground-breaking work on his "Roman cement" composite in the 1780s and 1790s, which despite its cheapness in manufacture and timely appearance at the end of the eighteenth century was no competition for the Coade stone artefacts which were then dominating the market with a selective clientele. At around this time, James Pulham I (1793–1838) was employed by William Lockwood & Co. in Woodbridge, Suffolk in manufacturing a cementitious artificial stone based on a type of Portland cement in the late 1820s and his son, James Pulham II (1820–1898), carried on and developed this theme more successfully in Broxbourne, advertising his Pulhamite Rockwork System as "*the building up on natural stone in complete imitation of a portion of a rocky cliff with unstratified stone or rock and bound where necessary with Pulham's cement made of the same colour and texture as lime or sandstone and tufa. It is recommended for substitution where no real stone or rock exists at or near and too expensive to get*". This idea certainly contributed to several large garden landscaping projects in the early nineteenth century where rock faces, stream diversions and waterfall construction were envisaged and it involved the marriage of natural rocks and artificial cementitious stone to create large aquiferous designs and grottoes (Riching & Lilly, *Rock Landscapes: The Pulham Legacy -Rock Gardens, Grottoes, Ferneries, Follies, Fountains and Garden Ornaments* 2012).

Undoubtedly, the major advent on the scene of cementitious artificial stone manufacture (Swann, *The Building Conservation Directory: 19th Century Cementitious Sculptures and Rockworks,* 2019) was made by Felix Austin, who in 1814 advertised a composition of his own invention from his premises in New Road, Rotherhithe, London, in which Roman cement formed the chief ingredient. In June 1827 he was operating from premises in Albany Street, Regent's Park, and then the following year in 1828 he bought the moulds and stock from the bankrupt business of a minor competitor of Eleanor Coade, namely Van Spangen, Powell & Co. of Bow, who used a Roman cement composite in their manufactured products. Austin some years later switched to Atkinson's cement and this was about twenty years after Aspdin's patent on Portland cement had been granted, and the company then retained this for later developments. Atkinson's cement was also known as Yorkshire or Mulgrave cement from its sandy raw material obtained from Sand's End in Yorkshire which was grey in colour, so avoiding the unfavourable brown and rather restrictive colour that the cementitious stone composite possessed through the use of

Parker's Roman cement. Austin's mortar mix was termed "puddingstone" as it contained crushed Portland stone (with grains between 1 and 10 mm in size) and he used iron armatures wrapped in zinc sheet to join the cast pieces together and strengthen the artefacts during their construction. He also advocated the incorporation of bricks and brown coloured Roman cement into the interior of his cast pieces to economise upon his more expensive cement which was then applied as an outer layer. The problem with Austin's statues is that upon exposure to a damp environment the iron rods would corrode from the inside and the mechanical stress induced would cause cracks to appear in the external cement structures: this was abetted and exacerbated by the mortar shrinkage, especially if this had been made with a high water content, a calcite leaching out process and problems associated with local regions of high sulfate content in the mortar mixture. Hence, we would anticipate that Austin's artefacts would not be as resistant to deterioration in damp, wet or polluted environments as would Coade stone. Austin was joined in 1842 by John Seeley, whose name coincidentally sounds very similar to that of the John Sealy who partnered Eleanor Coade in the late eighteenth century and who had died in 1813. Austin & Seeley, as the firm then became known, went on to become major players in the field of cementitious artificial stone, particularly for external garden statuary and they produced a catalogue advertising their "*Collection of Ornaments for Gardens, Parks and Pleasure Grounds*"; their success was determined largely, as was Eleanor Coade's some seventy years hitherto, by their association with important designers and landscape architects in the mid to late 19th Century. Austin & Seeley placed a circular in *The Ecclesiologist* stating "*Austin and Seeley beg to add that their work is executed without the application of heat*" (*The Ecclesiologist* 1842). From this statement we can conjecture that some comments had been made publicly that perhaps they were merely reviving and remarketing the Coade stone process for their own artefacts and not venturing into the new area of cementitious artificial stone. An example of an Austin & Seeley eagle statue which shows evidence of some internal corrosion and deterioration manifest visibly by a cracking of the leg of the statue is shown here and will be discussed later. The conservation of such a piece is not a simple task as the mere filling of the crack with a proprietary cement will not remove the problem of corrosion damage ongoing to the interior of the artefact and it is necessary to cut open the affected part and remove the corroded iron strut or armature that is the origin of the damage. Austin died in 1850 and then John Seeley bought the business and ran it successfully until he died in 1870: the stock was then sold off over the following years between 1871 and 1877.

2.3 Successors to Coade & Co. for Ceramic Artificial Stone

In 1843, Routledge & Greenwood (also called Routledge & Co.), who had taken over the failed Coade stone business from Thomas Croggon in 1837, sold off the moulds and factory hardware, much of which was bought by Mark Blanchard, who had previously worked at the Coade factory, and by John Blashfield. At the age of 23, Blanchard set up in Blackfriars Road, London, to manufacture terracotta and ornaments. He also made scagliola, a type of artificial coloured marble made from a composite plaster on a wooden former and he still used Coade's name, in fact, his first catalogue (Fig. 2.5) is a copy of the Coade's Gallery card designed by John Bacon (Fig. 2.3) even reproducing the identical montage of figures and claiming his succession to Eleanor Coade: there are even several statutes known to be of Blanchard origin that are made from Coade moulds which bear the stamp of Coade & Sealy, Lambeth. His stonewares were a vitrified ceramic reminiscent of Coade stone and his known examples have survived very well over the years. In the 1851 Great Exhibition at the Crystal Palace, Blanchard won a prize for his ceramic artificial stone artefacts (Fig. 2.6) and this attracted the attention of John Marriott Blashfield. In 1845, Blashfield had bought the business of Wyatt, Parker & Co. of Poplar, London, which had been set up to develop the patent of the Rev. James Parker for his Roman cement artificial stone composite in 1796: he ran the factory and in 1859 using their moulds he advertised the largest gallery of specimen statuary to be accumulated since that of Eleanor Coade (Fig. 2.7). John Blashfield wrote an account of artificial stone and its uses for the manufacture of artefacts in 1855 (Blashfield, *An Account of the History and Manufacture of Ancient and Modern Terracotta and its Use in Architecture as a Durable and Elegant Material for Decoration,* 1855). In the interim he was involved with the Minton China Works in Staffordshire and in the production of scagliola, which was then much in demand in Victorian England. He sold his specialist ceramic stone to furnish the Houses of Parliament and the Tsar's Winter's Palace in St Petersburg. Becoming bankrupt later, Blashfield eventually moved out of London and set up in Stamford in 1859 under the name Blashfield, Stamford, producing a similar catalogue advertising over seven hundred items in artificial stone (Fig. 2.8) (Stratton, *The Terracotta Revival,* 1993; Tavenor-Parry 1913). Unlike Blanchard, Blashfield did not claim his artefacts were in succession to Coade stone and he reverted to the older terminology of *Terracotta* to describe his composite material. Perhaps Mark Blanchard (born in 1816, just five years before Eleanor Coade died) felt that he was truly a successor to Coade in that he was apprenticed at the Coade manufactory in Lambeth during its last years under the proprietorship of Thomas Croggon. Blanchard was made bankrupt again in 1872 and thereafter he rebadged his business as The Stamford Terracotta Co., where he concentrated upon making scagliola until the company was finally liquidated in 1875. Mark Blanchard died in 1892. Blashfield and Blanchard were both, therefore, successors to Eleanor Coade in the manufacture of high-temperature fired ceramic artificial stone, although they had arrived at this position via different routes, and this was still accomplished successfully against the ever increasing competition from the cold-cured cementitious stone manufacturers. Examples of the impressed marks used by Mark Blanchard and John Blashfield on their artificial stone artefacts are shown in Figs. 2.9 and 2.10, respectively. A script mark used by John Blashfield is shown in Fig. 2.11.

Fig. 2.5 Catalogue of M.H. Blanchard, Sons, and Co., Successors to Coade's original works, Established 1769. Blackfriars Road, London, SE. Not dated but probably before the Great Exhibition of 1851, where they won a Prize Medal: the later catalogue of 1869, shown below in Fig. 2.6. Depicts this award. This earlier Catalogue clearly claims a succession from Coade & Co. as it shows the same John Bacon design as that depicted on the Eleanor Coade Gallery Card of 1799 and is accompanied by the following statements in typical Coade style: "*The Attempts of Time to Destroy Sculpture and Architecture Defeated by the Vitrifying Aid of Fire*" and "*The Power of Fire shall O'er thy Strength Prevail Till Thou and Fire with this Greta Globe shall Fail*"

Fig. 2.6 Later catalogue of Blanchard, Sons and Co., issued in 1869, an illustrated catalogue of beautiful works of art, showing the Prize Award Medal from the Great Exhibition of 1851

2.3.1 Other Types of Artificial Stone

Other types of artificial stone used variants of the cement and concrete cold cure recipes to make casts from wooden or iron moulds which could technically be accomplished at or near their place of usage, thereby economising on transport costs. William Ranger (1799–1863) patented an artificial stone manufacturing process to simulate Portland stone in 1832 and 1834 which used a weak hydraulic quicklime and a hot water mixture which proved suitable for the manufacture of capstones and ashlar blocks. Later in the nineteenth century, pigmented stone came into vogue and William Lascelles patented his "cement slab" in 1875 which could be coloured as required to match in with or act as a contrast to other natural stone masonry, especially Portland stone. In 1844, Frederick Ransome created a patent siliceous stone which comprised sand and powdered flint mixed in an alkaline solution which was then processed by heating in a steam bath and could be moulded on site into slabs and other architectural features. A similar siliceous stone was *Victoria stone* which comprised three parts of Mountsorrel (Leicestershire) granite and one part Portland cement which was mixed mechanically and cast into moulds.

No. 232.
BUST OF HER MAJESTY THE QUEEN, FROM A BRONZE BY WEIGALL.
Height, 14 inches.

MANUFACTURED BY J. M. BLASHFIELD
STAMFORD, Lincolnshire.

A Catalogue
OF
SEVEN HUNDRED ARTICLES,
MADE IN
Patent Terra Cotta,
AND
RED AND CANE-COLOURED POTTERY,
BY
J. M. BLASHFIELD,
STAMFORD, LINCOLNSHIRE.

London Depôt, 16, Great Marlborough Street. W.

The entire Catalogue will comprize nearly Three Thousand examples of subjects connected with Sculpture and Architecture, and for which the Patentee has Models.

Fig. 2.7 Earlier catalogue of J.M. Blashfield London Depot, 16 Great Marlborough Street, W. A Catalogue of seven hundred articles made in patent Terra Cotta and Red and Cane-coloured pottery by J.M. Blashfield and manufactured by J.M. Blashfield, Stamford, Lincolnshire

The casts were then placed in a sodium silicate solution for two weeks to harden them. The idea of a moulded siliceous stone was not a novel one created in the late nineteenth century: as early as 1821 William Weston Young had conceived the idea of manufacturing silica bricks using a particularly rich silica outcrop (comprised of >92% silica) at Dinas Rock in the Vale of Neath, which he treated with a lime flux and cast into moulded fire bricks which then underwent thermal processing at the relatively very high temperature of 1400 °C, initially carried out at the Nantgarw China Works site in Glamorgan using their high-grade anthracite coal-fired porcelain kiln. These *Dinas bricks*, as they came to be known, were ideal for the heavy industrial iron and steel works of the Industrial Revolution, where the kilns had previously suffered serious mechanical and chemical thermal damage through their degradation at the high operating temperatures required to smelt molten iron and steel (Edwards, *Porcelain to Silica Bricks: The Extreme Ceramics of William Weston Young, 1776–1847*, 2019). However, although Young's main interest primarily was in reviving the fortunes of the Nantgarw soft paste porcelain manufactory, which had finished production of its quality phosphatic porcelain in 1820, there was no suggestion that he was intent upon replacing this household porcelain with garden ornamentation or building constructional artefacts although he did make several

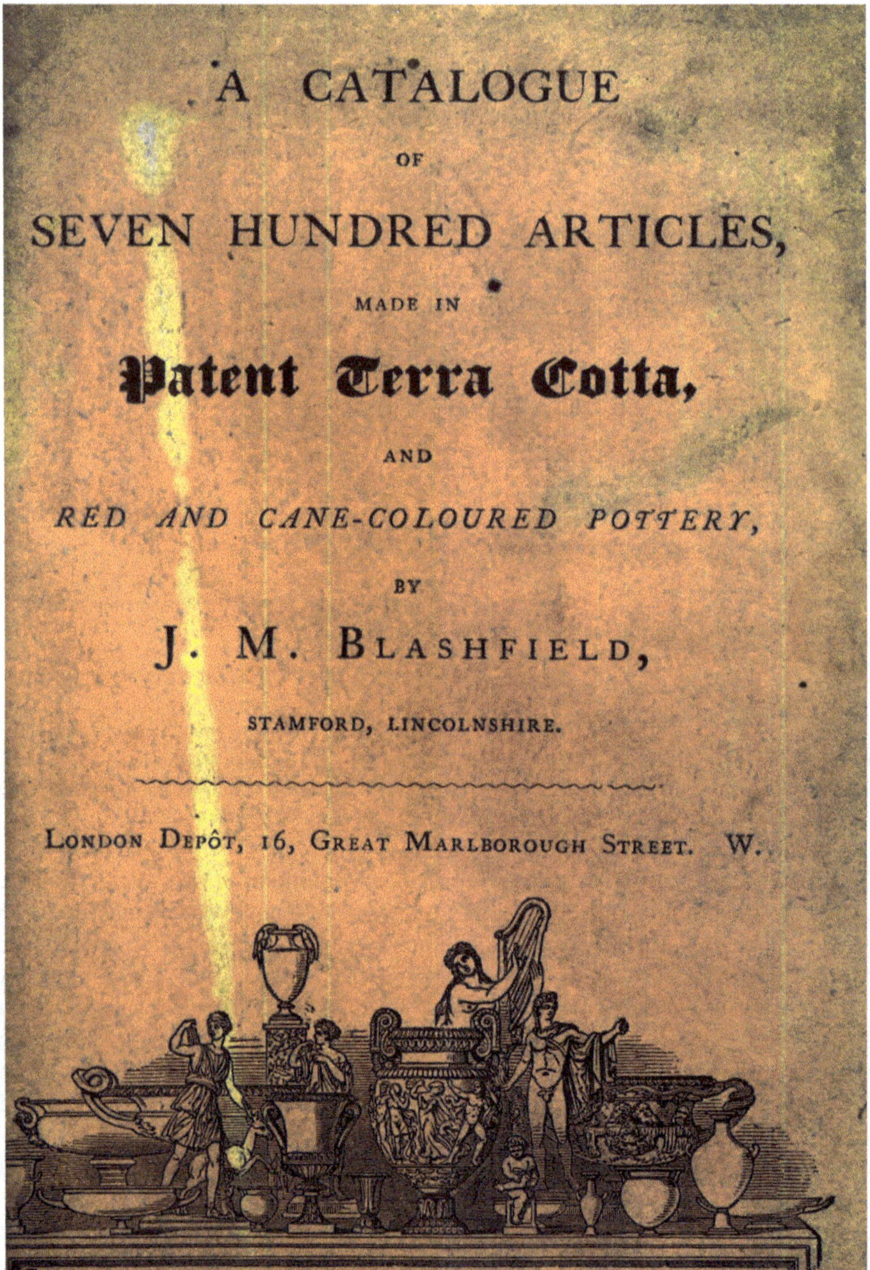
A CATALOGUE
OF
SEVEN HUNDRED ARTICLES,
MADE IN
Patent Terra Cotta,
AND
RED AND CANE-COLOURED POTTERY,
BY
J. M. BLASHFIELD,
STAMFORD, LINCOLNSHIRE.
LONDON DEPÔT, 16, GREAT MARLBOROUGH STREET. W.

Fig. 2.8 Later Blashfield catalogue with identical wording as that in Fig. 2.7

Fig. 2.9 Impressed mark of Blanchard artificial stone — TERRA COTTA M.H. BLANCHARD & CO Blackfriars Road London. Undated

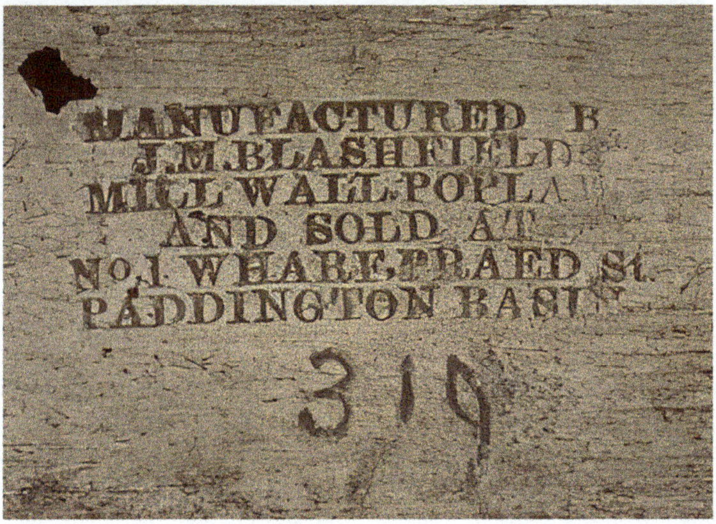

Fig. 2.10 Impressed mark of Blashfield artificial stone—MANUFACTURED BY J M BLASHFIELD, MILL WALL POPLAR AND SOLD AT N$^{o.}$ 1 WHARF PRAED St. PADDINGTON BASIN. Undated

partially successful attempts to create a Nantgarw hard paste porcelain in the same kilns, which did not receive support from a clientele that was used to the perceived superior soft paste Nantgarw porcelain (Colomban et al. 2020).

2.4 Facts and Fiction Relating to Coade Stone and Other Artificial Stones

In common with other ceramic historical research from the eighteenth century, the background studies of Coade stone have engendered many statements historically that upon careful checking reveal some misconceptions that have originated from

Fig. 2.11 Script signature of J.M. Blashfield in black pigment. Undated

early historians making some unsupportable conclusions, generally on the basis of hearsay or an incorrect interpretation of documentary evidence, which have been made without the evidential support that is now demanded in a modern forensic holistic approach to ceramics research. In other words, opinions that were generally held at the time, possibly made between one hundred and two hundred years ago, have been incorporated as raw facts into a caucus of literature which has not been examined forensically or holistically until now, whereby analytical scientific information and historical documentation have been properly evaluated together to weed out assertions that have been made without substance or real significance. These myths and legends relating to Coade stone will be discussed and itemised later in this book.

References

S.J. Bernard Burke, *Burke's General Armory—The General Armory of England, Scotland, Ireland and Wales: Comprising a Registry of Armorial Bearings from the Earliest to the Present Time* (Harrison & Sons, London, 1884)

J.M. Blashfield, *An Account of the History and Manufacture of Ancient and Modern Terracotta and its Use in Architecture as a Durable and Elegant Material for Decoration* (John Weale, London, 1855)

A.T. Bolton, *The Architecture of Robert and James Adam* (Country Life, London, 1922)

E. Coade, J. Sealy, *A Descriptive Catalogue of Coade's Artificial Stone Manufactory at King's Arms Stairs, Narrow Wall, Lambeth: Opposite Whitehall Stairs* (With Prices Affixed, Manufactory and J. Strahan, London, 1784)

P. Colomban, H.G.M. Edwards, C. Fountain, Raman spectroscopic and SEM/EDXS analysis of highly translucent Nantgarw soft paste porcelain. J. Eur. Ceramic Soc. **40**, 4664–4675 (2020)

L.W. Dillwyn, Notes of Recipes for Porcelain Bodies and Glazes 1815–1817, Library of the Victoria & Albert Museum, South Kensington, London, deposited by John Campbell Esq. in 1920. Reproduced in Eccles and Rackham, *Analysed Specimens of English Porcelain in the Victoria and Albert Museum Collection*, see reference below

H. Eccles, B. Rackham, *Analysed Specimens of English Porcelain in the Victoria & Albert Museum Collection* (Victoria & Albert Museum Publications, London, 1922)

H.G.M. Edwards, *Porcelain to Silica Bricks: The Extreme Ceramics of William Weston Young, 1776–1847* (Springer -Nature, Dordrecht, 2019)

H.G.M. Edwards, *Porcelain Analysis and Its Role in the Forensic Attribution of Ceramic Specimens* (Springer -Nature, Dordrecht, 2021)

H.G.M. Edwards, *Welsh Armorial Porcelains: Nantgarw and Swansea Crested China* (Springer -Nature, Dordrecht, 2022)

H.G.M. Edwards, M.R.D. Seaward, Raman spectroscopy and lichen biodeterioration. Spectrosc. Eur. **5**(1), 16–20 (1993)

H.G.M. Edwards, D.W. Farwell, M.R.D. Seaward, C. Giacobini, Preliminary Raman microscopic analyses of a lichen encrustation involved in the biodeterioration of renaissance frescoes in Central Italy. Int. Biodeterior. **27**, 1–9 (1991)

H.G.M. Edwards, D.W. Farwell, R. Jenkins, M.R.D. Seaward, Vibrational Raman spectroscopic studies of calcium oxalate monohydrate and Dihydrate in lichen encrustations on renaissance frescoes. J. Raman Spectrosc. **23**, 185–189 (1992)

K. Esdaile, Coade stone. Archit. Build. News **CLXI**(19th January Issue), 94–96 (1940a)

K. Esdaile, Coade stone. Archit. Build. News **CLXI**(26th January Issue), 112–114 (1940b)

Etchings of Coade's Artificial Stone Manufactory at King's Arms Stairs, Narrow Wall, Lambeth, 1784, British Library

J. Fairbairn, *Crests of the Families of Great Britain and Ireland and Their Kindred in Other Lands, Volumes I and II* (T.C. and E.C. Jack/New Orchard Publishing, Edinburgh and London/London, 1905/1986)

C. Fowler, Some remarks on Terracotta and artificial stone as connected with architecture. Civil Eng. Archit. J. **13**, 215–216 (1850)

A.J. Francis, *The Cement Industry, 1796–1914: A History* (David & Charles, Newton Abbott, 1977)

F.H. Hansford, Eleanor Coade of Lyme Regis. The Dorset Year Book (1960–1961), p. 27. *Proceedings of the Institute of Civil Engineers: Structures and Buildings*, Volume **116/3**, 1996

A. Kelly, An expensive present: the Adam screen in Rio de Janeiro. Burlingt. Mag. **126**(September Issue), 548 (1984)

A. Kelly, *Mrs Coade's Stone* (Self-Publishing Association, Worcestershire, 1990)

D. Pincot, *An Essay on the Origin, Nature, Uses and Properties of Artificial Stone* (Richard Hett, London, 1770)

C. Riching, J. Lilly, *Rock Landscapes: The Pulham Legacy -Rock Gardens, Grottoes, Ferneries, Follies, Fountains and Garden Ornaments* (Garden Art Press/NHBS, Devon, 2012)

H. Rix, S. Emmitt, Issues of authenticity when conserving historic imitative crafts. Archit. Conserv. **29**(1), 1–19 (2023)

G. Roberts, *The History and Antiquities of the Borough of Lyme Regis and Charmouth* (Samuel Bagster & Sons, London, 1834)

S. Scott, *Artificial Stone: A Successful Substitute for Natural Stone?* (Haddonstone Ltd, Northampton, 2009)

C. Stanford, Revisiting the origins of Coade stone. Georgian Group J. **XXIV**, 95–116 (2016)

M. Stratton, *The Terracotta Revival* (Victor Gollancz, London, 1993)

S. Swann, *The Building Conservation Directory: 19th Century Cementitious Sculptures and Rockworks* (Cathedral Communications Ltd, 2019)

J. Tavenor-Parry, An episode in the history of English Terracotta. Archit. Rev. **XXXIII**, 119–122 (1913)

The Ecclesiologist, **Volume 1**, No. VIII, May 1842, 27 (1842)
P. Thomason, Coade Stone Restoration. Cudworth, Dorset, personal communication
N. Valpy, A. Kelly, Advertisements for artificial stone. Trans. Engl. Ceram. Circ. **12**, 206–226 (1986)
F.R.S. Yorke, Concrete and artificial stone. Archit. J. **75**, 729 (1932)

Chapter 3
Artificial Stone: The Precursors, Contemporaries and Later Variations of Coade Stone

Abstract Further development of the early precursors to Coade artificial stone from the sixteenth century through John Dwight's Fulham Pottery in the 1690s manufacturing high temperature fired porcelain and refractory stonewares, through the early patent of John Holt in 1722 to the wares of Blanchard and Blashfield in the late nineteenth century. Discussion of fired clays and of hydraulic and nonhydraulic lime-based cements, such as Roman cement, and notable landmarks including John Smeaton and the Eddystone Lighthouse.

Keywords Oxnead · John Holt · John Dwight · Fulham pottery · Cologne stonewares · Hydraulic and nonhydraulic lime cements · Roman cement · Pozzolanic cement

3.1 Blashfield and Blanchard

Many historical accounts of Coade stone accredit Eleanor Coade with the invention of artificial stone: this is not correct, although there is no doubt that her astute business acumen, flair for design, her engagement with the most renowned sculptors and landscape architects of her day and a knowledge of what her clients wanted in the way of neo-classical statuary and building decorations materially supported her business portfolio enhancement and also provided the launchpad for several contemporaries and later ventures in this area. However, the precursors of her Coade artificial stone artefacts are not so easily quantifiable. The first patent recorded for an artificial stone was not filed until 1722, when Richard Holt with Thomas Ripley laid claim to their synthesis of a liquid stone ceramic that could be moulded, then fired and which had a remarkable resistance to erosion in an inclement environment that far surpassed the properties of natural stone, particularly the popular building material, Portland limestone. There had certainly been earlier attempts at creating such an artificial synthetic material, which had their origins in the manufacture of fired household ceramics, such as terracottas, stonewares and porcelains, but the usage of these materials for architectural purposes required a further quantum leap forwards. The first thing that is noted by a reader on the topic of Coade stone generally is that is frequently referred to as *terracotta*: in fact, John Marriott

H. G. M. Edwards, C. J. Brooke, *Coade Stone*, Cultural Heritage Science,
https://doi.org/10.1007/978-3-031-71432-0_3

Blashfield, who ran a successful business in artificial stone ornaments from the 1850s, called his products "terracotta" (Blashfield, *A Catalogue of Seven Hundred Articles made in Patent Terra Cotta, and Red and Cane-Coloured Pottery,* 1882) and the Blashfield stone composition is indeed believed to be very closely similar to that of Coade stone. Blashfield was one of several successors to Eleanor Coade who carried on with the artificial stone business after Coade & Co. had closed down in the late 1830s: in 1851, the Great Exhibition in the Crystal Palace set up a separate classification for terracotta and artificial stone exhibits for their manufacturers. In 1845, Blashfield, who had worked at Minton's in Staffordshire on tiles and *scagliola*, bought into Wyatt Parker in London and then took over the manufactory; it is believed by some historians that he purchased all of the moulds from Eleanor Coade's factory at its closure, this may be true, but he actually collected hundreds of casts of Coade stone ornaments "*with the kind liberality of some of my earliest patrons*", thereby acquiring the largest gallery of artificial stone specimens seen since the dispersal of the Coade collection in the late 1830s, as reported in the *Stamford Mercury* of February 18th, 1859. His advertisement for his Gallery clearly indicates that he claims to be the natural successor for ornaments and artefacts of Coade stone and that he has continued with the designs as sculpted by John Bacon for Eleanor Coade. The advertisement catalogue for his artificial stone artefacts shows the Coade kiln, even displaying its classical motto above the entrance, and is accompanied by his statement, which echoes sentiments reminiscent of those that were expressed by Eleanor Coade:

> The Attempts of Time to Destroy Sculpture and Architecture Defeated by the Vitrifying Aid of Fire . . . .above the triplet . . .
>
> In Vain thy Threats O Time these Arts Assail
> The Power of Fire shall o'er thy Strength Prevail"
> Till Thou and Fire with this Great Globe shall Fail.

Blashfield's works and showrooms were at Blackfriars Road, London. He moved out of London in the late 1850s to Stamford in Lincolnshire because of impending bankruptcy over some inappropriate property dealings but he still maintained his artificial stone business and had some contracts with Burghley House nearby. A classic example of Blashfield's adoption of Eleanor Coade's mantra is his balustrading in "Coade stone" in the gardens of Castle Ashby, the ancestral home of the Marquess of Northampton, and this is shown in Fig. 3.1. Castle Ashby in Northampton was built in 1564 and was landscaped in 1761 by Lancelot "Capability" Brown. In 1862, Theodosia, the Marchioness of Northampton, engaged Sir Matthew Digby Wyatt and W.B. Thomas to redesign the formal gardens at Castle Ashby and they made a lavish use of Blashfield's artificial stone on the house itself and in the gardens, which were completed in 1864. A most interesting balustrade in the garden was further added a little later in 1866, in which the balusters were replaced with letters which spelled out the Biblical saying from Christ's Sermon on the Mount as iterated in Luke 12: 27:

> CONSIDER THE LILIES OF THE FIELD HOW THEY GROW; THEY TOIL NOT NEITHER DO THEY SPIN **BUT YET I SAY UNTO YOU** THAT SOLOMON IN ALL HIS GLORY WAS NOT ARRAYED LIKE ONE OF THEM

Fig. 3.1 Section of artificial stone balustrade manufactured by John Marriott Blashfield and erected in 1866 in the formal gardens of Castle Ashby, Northampton, family seat of the Marquis of Northampton designed by Sir Digby Wyatt. Unusually, the balusters have been replaced by letters which spell out a Biblical phrase

The section of balustrading shown in Fig. 3.1 reproduces the part highlighted in bold script above. Theodosia never saw her biblical balustrade finished as she died in 1865 prior to its completion and a Latin dedication to her was added later to her balustrade.

Mark Blanchard, Son & Co. also claimed a succession from Eleanor Coade and as "Terracotta Manufacturers to The Royal Family and the Several Departments of Government"; he too published a catalogue of Coade type stone artefacts, rather later than Blashfield in 1869, but he too moved out of London in the early 1870s to Bishops Waltham in Hampshire.

The definition of terracotta (Latin: *terra cocta*, fired/cooked earth) is that it is a clay—based ceramic which has been fired at a temperature of between 600 and 1000 °C, which may be glazed or unglazed, where the body is porous and ranging in colour from ochre to deep red. Archaeologists additionally tend to refer to terracotta artefacts as those which have not been made on a potter's wheel, so the term is then rather divisive in definition of the type of object rather than its material composition or colour. Another differentiation defines terracotta as pieces made for the construction industry rather than for the household table, which then retain the alternative term, earthenwares. A particular distinction that is reflected in the latter definition is that terracottas are not usually glazed, and if so they are then referred to as faience, as thereby they are rendered non-porous for household use with liquids and for cooking and storage vessels in the home. Faience itself is split into two main categories, dependent upon whether both sides of a utilitarian implement are glazed (delftwares) or not (majolica /maiolica). A further category of faience is provided by the Asian ceramic celadons, especially those from China, Korea and Vietnam, which comprise pottery artefacts that are accompanied by a vivid green glaze added in simulation of jade. Hence, the definition of terracotta is perhaps non-specific and rather all-embracing but most people believe that it really refers to the characteristic reddish-brown colour that is imparted to the fired pottery body by the iron (III) oxide, Fe_2O_3, present in the sand and clays that are used as a raw material. In the sixteenth century and thereafter, the deliberate selection of white clays (for example, kaolinite) and fine white river sand minimised the iron oxide content, with usually the addition of a feldspar or china stone to the raw materials, and the fired bodies that resulted were a pleasant white or greyish-white in appearance: this gave rise to the much admired versions of stonewares, and especially "Cologne ware", which were fired at higher temperatures (between 1150 and 1300 °C) than terracotta, that could be decorated with a stylus inscribing the wet paste after moulding and before firing, and which were often coloured superficially with a salt glaze. Such wares, generally termed stonewares, or in England "ironstone", in the nineteenth century were manufactured by several well-known companies such as Royal Doulton, Wedgwood and Miles Mason. Cologne wares achieved their name (alternatively described as *gres de Flandres*) from their centres of manufacture based near Cologne (Nordrhein-Westfalen) and their export from that city. An interesting variation on the meaning of "terracotta" is provided by Arthur Fontaine in his recent search for Coade stone artefacts in Germany (Fontaine, *Der Englisch Coade -Stein—die Villeroy & Boch "Terakotta"*, 2019), which appear to be rather rare, and he alleges that Villeroy and Boch in the mid-nineteenth century re-worked Coade stone from the published recipe into garden ornaments which were sold under the name of *terracotta*. Fontaine also has discovered that Martin Klauer in Weimar and Friedrich Schinkel in Berlin also attempted to break into this market by manufacturing versions of Coade stone there.

3.2 The Fulham Pottery and John Dwight

The most intriguing precursor of Coade artificial stone in 1769, of Daniel Pincot's in 1767 and prior to that, Richard Holt's patented version of a clay-less fired artificial stone in 1722 must be considered to be that of John Dwight at the Fulham Pottery in the late seventeenth century. Although never specifically termed artificial stone, Dwight's fired ceramic has been called a *porcellaneous stoneware* (Sir Arthur Church 1890): a term that seems clearly self-contradictory on closer inspection as the major criterion for the differentiation between fired porcelains and fired stoneware ceramics is that the former exhibits translucency and the latter does not! However, it is apparent from reading accounts of Dwight's manufactured artefacts that historians cannot decide if he actually made porcelain or not, hence the ubiquitous and rather indefinite term *porcellaneous stoneware,* which was applied and has remained to describe this type of artefact. John Dwight was born in Todenham, Gloucestershire, between 1633 and 1636 and read natural philosophy at Oxford University, becoming a laboratory assistant to Robert Boyle, the "Father of Chemistry" (*Authors*: this title has also been bestowed upon the French chemist, Antoine Lavoisier, who lived some 100 years later). Dwight became diocesan registrar to Bishop John Wilkins in Chester but in 1670 he established the Fulham Pottery in

Fig. 3.2 Kiln at site of the Fulham Pottery, London, stoneware manufactory of John Dwight from 1670 until 1703, excavated by the Museum of London

New King's Road, London, to manufacture high quality stoneware in competition with the imports from Cologne (Peart 2009).

Fulham Pottery was situated at the junction of the New King's Road and Burlington Road, not far from Putney Bridge on the River Thames, with the express purpose of manufacturing high quality salt-glazed stoneware which at that time was the preserve of émigré Dutch artisans. Recent archaeological excavations undertaken at the site of the Fulham Pottery by the Museum of London have revealed the presence of a ceramic kiln which has now been preserved on this urban site (Fig. 3.2). Upon the foundation of his pottery at Fulham, John Dwight filed a patent in April 1671 describing his experiments in the use of clays and minerals to make a high quality "stoneware" and a phrase from this patent is particularly revealing, namely:

> . . . the discovery of the mistery of **transparent earthenware** commonly known by the name of **porcelaine** in China.

The wording of this patent is critical in that Dwight has been credited by several historians with the manufacture of only stonewares at Fulham and the granting of this patent for the manufacture of a "*transparent earthenware*," i.e. "*porcelain*", rather than just the opaque so-called brown salt-glazed earthenware, is revealing, although historians have explained this discrepancy by assuming that his earthenwares were themselves "transparent" which surely is a specious and unsubstantiated explanation as earthenwares are defined as being opaque to transmitted light! Sir Arthur Church had some detailed knowledge of Dwight's patent and he alleged that Dwight must have achieved the manufacture of something very similar to what we now call porcelain in that "*... he produced if not porcelain then something distinctly porcellaneous*" (Church, *English Porcelain: A Handbook to the China Made in England During the 18th Century as Illustrated by Specimens Chiefly in the National Collection*, 1885 and 1894). The substitution of the adjective *porcellaneous* in place of the definitive noun *porcelain* is really addressing a matter of degree, whilst nevertheless reflecting the quality and extent of the transparency of the material and surely must differentiate Dwight's artefacts and material thereby from earthenware and stoneware in at least being somewhat transparent or translucent.

3.2.1 The Royal Society and Ceramics Research in the Seventeenth and Eighteenth Centuries

The role of the Royal Society of London in English ceramics research and manufacture has been rather played down to date, although it was mentioned in an earlier ceramics work by Solon (*The Old English Potter,* 1883). Daniels (*The Origin and Development of Bow Porcelain, 1730–1747, Including the Participation of the Royal Society, Andrew Duche and the American Contribution,* 2007) has reopened the history of the Royal Society's scientific interest in the monitoring of John Dwight's experiments with her interpretation of several key documents (Haselgrove & Murray, *John Dwight's Fulham Pottery, 1672–1978: A Collection of Documentary Sources*, 1979) which demonstrate clearly that in the mid- to late-seventeenth century, within a few years after its foundation during the Restoration of King Charles II in 1660, the Royal Society was actively encouraging and sponsoring research and experimentation in refractory materials with John Dwight at the Fulham Pottery and that this was occurring some 40 years prior to the successful Meissen porcelain manufacturing operation undertaken by King Augustus, Elector of Saxony. Because of the key scientific link between John Dwight and Robert Boyle, Fellows of the Royal Society were monitoring closely the high temperature synthetic ceramics work being carried out at the Fulham Pottery in the 1670s. The eminent scientist Robert Hooke FRS, described as the "Leonardo da Vinci" of English science, on February 1674 (in Haselgrove & Murray, *John Dwight's Fulham Pottery, 1672–1978: A Collection of Documentary Sources*, 1979) confirmed the successful production by Dwight of his "*English china*" including some figures and:

> Severall little Jars of severall colours all exceeding hard as flint, very light and very good shape.

And again on May 16th, 1674, he described Dwight's products as

> In glazed with ashes. Very hard and close excessive deer.

The latter comment perhaps refers to Dwight's potential use of a silicaceous-alumina-calcareous (SiAlCa) glaze with perhaps some lead oxide present in addition. The role of someone like Robert Hooke taking an interest in porcelain manufacture should not be underestimated. In Stephen Inwood's biography it was said of Hooke by his contemporaries that "*Hee makes questions to those he knows are skillfull in them*" (Inwood, *The Man Who Knew Too Much*: *The Strange and Inventive Life of Robert Hooke, 1635–1703,* 2002). In other words, Hooke typified the person who knows what question to ask and from whom he will obtain the best information in return. Knowledge was eagerly sought after, traded and exchanged in the *Age of Enlightenment* and in seventeenth century London the best sources of practical knowledge were to be found in the coffee houses and taverns and Hooke is recorded as saying that between 1672 and 1680, the active years of the Dwight Fulham Pottery porcelain venture, he personally visited 60 different coffee houses and 80 taverns. London's first coffee shop, Pasqua Rosee's Head, in St Michael's Alley off Cornhill, opened in 1652; by 1663, there were 82 coffee shops in the capital and by 1703, the year of John Dwight's death, there were 500 of them (Long, *At the Sign of Atlas: The Life and Works of Joseph Moxon FRS, A Restoration Polymath,* 2013, p. 89), which gives an idea of the valuable conduit of information that could be accessed by the patron savants, fabricants and luminaries of the time (Edwards, *Porcelain Analysis and Its Role in in the Forensic Attribution of Ceramic Specimens*, 2022, Appendix VIII). Caroline Stanford has recently explored the idea that Robert Hooke and John Dwight were pioneers in the adoption of artificial stone (Stanford 2020).

Dwight experienced problems with the glazing of this *white porcellane* stoneware body, which was suspected by Ramsay & Ramsay (*The Evolution and Compositional Development of English Porcelains from the 16th Century to Lund's Bristol c. 1750 and Worcester c. 1752- the Golden Chain",* 2017) to arise from his adoption of a silica -alumina -calcareous glaze composition which required very high firing temperatures, which Dwight was unable to achieve reliably in his kiln. This had also been suggested by Tite et al. (1986), who attributed the defects in Dwight's manufactured products to the adoption of an intractable SiAlCa glaze which had problems in adhesion to the ceramic body substrate.

Dr. John Houghton FRS in 1693 (Haselgrove & Murray, *John Dwight's Fulham Pottery, 1672–1978: A Collection of Documentary Sources,* 1979) maintained that "*Mr Dwight of Fulham uses clay, the same earth China ware is made from and this is made by not lying long in the earth but in the fire . . . . we make as good China here as any in the world*". Houghton wrote again in 1694, bemoaning the fact that a restricted importation was in effect for the hard paste porcelain from China of "*curious manufacture*", but he sought solace in the fact that "*Mr Dwight of Fulham has done it, casting thin and firing it at high temperature*".

Sir John Lowther FRS in 1698 (Haselgrove & Murray, *John Dwight's Fulham Pottery, 1672–1978: A Collection of Documentary Sources,* 1979) visited Dwight at his Fulham Pottery and was shown "*20 or 30 varieties of more china like than is in ye world beside, nothing in Germany is like his, nor had he any help from them at setting up, but owes all to his own studies*". The reference to Germany has caused some commentators to assume that this remark was made in comparison of Dwight's synthetic stoneware to the *Cologne ware* that was being imported from Germany—but the constant reference to translucent porcelain wares admixed with these fired glazed earthenwares has caused much confusion amongst historians in the interpretation of what materials were actually being produced at Fulham and the answer could simply be that in fact both were being made there!

The only chemical analysis that has been carried out thus far into the Fulham Pottery products is described in the comprehensive work of Tite et al. (1986), which is entitled "*A technological study of Fulham stonewares*". The title of this paper could be misinterpreted as it seems to exclude specifically any porcelains made at the site but the word *stonewares* should be regarded in this instance as being all-encompassing in this respect as this analytical paper describes the study of high temperature fired shards that were found and analysed at the Fulham Pottery site, which included both stonewares and porcelain. In the 1970s, the Museum of London undertook an archaeological excavation of the Fulham Pottery site and recovered some shards of "brown and white stoneware" and also some coded test pieces of a *porcellaneous* material which matched the details given in Dwight's two notebooks which outlined his experimentation (Haselgrove & Murray, *John Dwight's Fulham Pottery, 1672–1978: A Collection of Documentary Sources*, 1979; Green, *John Dwight's Fulham Pottery: Excavations, 1971–1979*, 1999). The Museum have clearly stated that several of these pieces can be definitely classified now as porcelain and are translucent. Tite et al. (1986) analysed two shards of brown stoneware and five shards of a white "stoneware" along with three samples of a clay raw material found at the site using SEM/EDAXS and optical microscopy. It is interesting that the clay specimens were subjected to wet chemical analytical procedures involving extraction with hydrofluoric acid and fusion with lithium metaborate to determine the silica content and elemental percentages using atomic absorption spectroscopy, harking back to the early wet chemical digestion analyses of Eccles & Rackham (*Analysed Specimens of English Porcelain in the Victoria & Albert Museum Collection,* 1922). The clay raw material was confirmed as comprising a natural secondary ball clay, sourced from Poole in Dorset, with a mineralogical composition of 55–65% quartz, 15–25% mica and 15–25% kaolinite, in admixture with sand and lime as components rather than a naturally sourced raw material (Holdridge 1956). The brown and white stoneware shards were analysed using SEM/EDAXS and the five white porcelain shards relevant for this discussion produced the data shown in Table 6.1: both body and glaze were analysed independently and the glaze composition is very similar to that of the ceramic body in its major components except for a significantly larger contribution from the alkali metal oxides, which cumulatively comprise about 10% of the glaze mixture compared with about 4% for the ceramic body. Essentially, Dwight's porcellaneous stonewares are rather higher in silica and lower in alumina that Eleanor Coade's artificial stone but the other elemental oxides are present to a similar extent.

Tite et al. (1986) also inferred from their earlier optical vitrification measurements (Maniatis and Tite 1881) that a firing temperature of between 1200 and 1250 °C had been used for the white porcellaneous shards compared with 1100 °C for the lower grade brown stonewares. The white ceramic body analyses matched a selection of those quoted by John Dwight in his notebooks (Haselgrove & Murray, *John Dwight's Fulham Pottery, 1672–1978: A Collection of Documentary Sources*, 1979) which detailed seven variants of a white "transparent clay porcelain" body all comprising mixtures of 30–70% Dorset ball clay and a white earth/white sand component which was made separately by sintering quartz sand with nitre (potassium nitrate) and tartar emetic (potassium hydrogen tartrate), which effectively gave a potash frit of composition 97% silica and 3% potash. Of particular comparative interest in this analytical work was the observation that the fine white stoneware bodies analysed fell within the same compositional range of the *yinqing* Chinese porcelain bodies of the Yuan Dynasty (1260–1368) which had been analysed in a similar fashion by the same authors two years previously (Tite et al. 1984) but they are also significantly different in composition with the much lower alumina/silica ratios from the later Chinese porcelains of the Ming and Kangxi Dynasties. It is also interesting that the firing temperature adopted for the Yuan Chinese porcelains has been determined to be in the range 1200–1300 °C, which is a range analogous to that believed to have been used by John Dwight in his kilns at Fulham. The calculated mineral compositions of the Yuan and Fulham ceramic bodies are also quite similar with quartz 57–64% versus 57–73%, kaolinite 8–18% versus 19–30% and mica 14–27% versus 9–14%, respectively, for the Yuan and Dwight formulations. The very close similarity between Dwight's white "stoneware" and the *yinqing* Chinese porcelain is seen in the ternary compositional diagram, representing the extremes of $Na_2O + K_2O/Al_2O_3/SiO_2$ given in the article of Tite et al. (1986), where the data for both ceramic bodies spatially overlay each other, indicating that they had very similar relative elemental oxide compositions. It was also shown from an XRD analysis that the Dwight porcelain shards gave evidence of the presence of mullite and tridymite minerals, the key markers of a genuine *hard paste* porcelain which has been fired at a very high kiln temperature. This is surely well-defined analytical evidence that John Dwight had indeed produced a genuine porcelain as well as potentially his stonewares, which analytically is extremely similar to the Chinese *yinqing* porcelains of the Yuan Dynasty (1260–1368 CE), which preceded the Ming Dynasty and Qing Dynasty, and which together were responsible for the export of such significant quantities of Chinese porcelains into Europe in the eighteenth and nineteenth centuries. What Tite et al. (1986) also found is that the Dwight "stonewares", which perhaps can now be re-defined as really *proto-yinqing* porcelains, do not match the Kangxi porcelains of the early eighteenth century in composition, as verified later by Spataro et al. (2008, 2009).

Furthermore, Spataro et al. (2009) have discovered evidence for Dwight's reversion to a more amenable and achievable lower glaze firing temperature using a lead glaze, which he then used to recover several articles which had nevertheless successfully survived his preliminary but less successful SiAlCa glaze trials and kiln firing. It is frustrating that despite the observations of the Royal Society savants of the existence of Dwight's translucent porcelain, numbering certainly several dozen pieces or more which were seen by several independent observers in the last quarter of the seventeenth century, the location or survival of not one of these items can be identified with any certainty today. The Museum of London, as stated earlier, has undertaken an archaeological excavation of the Fulham Pottery site and have restored Dwight's kiln. During the excavations they unearthed several test fragments of fired porcelain bearing Dwight's code in a cobalt blue script, which were his so-called "coded test pieces" and which matched his experimental handwritten notes which were found earlier on the Fulham site by Lady Charlotte Schreiber in 1868, of which she meticulously made copies including Dwight's erasures. This was extremely fortunate as the original documents are now lost. The Museum of London confirm that these fired specimens were clearly "*porcelain*", for example, the specific test piece coded 97.90/29, measuring 33 mm by 27 mm. However, the Museum also notes that completed specimens were not found on site, only fragments, and they doubt whether any porcelain was actually manufactured for sale. The Museum of London also have re-constructed a vase from glazed fragments recovered during the excavations (Green, *John Dwight's Fulham Pottery: Excavations, 1971–1979*, 1999), which they now clearly display and have labelled as an example of John Dwight's *porcelain* alongside a similar unglazed vase reconstructed similarly from fragments found at the site.

3.2.2 *John Dwight—Sequel*

It can fairly be concluded, therefore, that sometime between 1672 and 1680, John Dwight had succeeded in manufacturing a ceramic body that was equivalent to the translucent hard paste porcelain perfected and manufactured in China in the Yuan Dynasty. It is true that he did experience severe problems with his glazing technique initially, which were partially overcome in later years by his institution of a re-glazing method involving a lead oxide flint glass additive, but it seems that the commercial exploitation of his new "refractory white stonewares" was never within his sphere of activity, and perhaps the scientific involvement of the Royal Society was behind this. Having shown that the creation of a highly refractory material for its potential adoption in crucibles was viable, the project had succeeded in its outcome and whether or not any crucibles were ever made with this new material is not known. Certainly, several high quality cast figurines were made in a high quality white stoneware of undefined composition, such as the poignant half-length recumbent figure of Lydia Dwight, John Dwight's daughter who died aged six years on March 6th, 1674, which is an example of one of the earliest European sculptures fashioned in ceramics (now in the Victoria & Albert Museum ceramics collection, number 1055–1871). An exhibition of some of Dwight's wares was made in 1992 (Horne, *John Dwight "The Master Potter" of Fulham, 1672–1703,* 1992) at Kensington Church Street.

The Fulham manufactory ceased to make "porcelain" and reverted to the manufacture of stonewares in 1703 after John Dwight had died. It is conjectural for some historians that he ever did make true porcelain at the Fulham Pottery but there is now much analytical evidence extant that he did so successfully. Statements from a range of contemporary observers confirm the highly translucent nature of his pieces, but there is little record of any commercial business aspect being undertaken in this respect (Page, *A History of the County of Middlesex: General—Ashford, East Bedfont with Hatton, Hampton with Hampton, Wick, Hanworth, Laleham, Littleton*, 1911). The influence of prominent members of the Royal Society in his manufacturing products, as noted above, centred upon their interest in the creation of refractory materials for crucibles able to withstand the highest temperatures achievable at that time for the extraction of metals from their mineral ores, which is a natural progression out of the alchemical principles from which the science of chemistry was born (Daniels, *The Origin and Development of Bow Porcelain, 1730–1747. Including the Participation of the Royal Society, Andrew Duche and the American Contribution,* 2007).

It is apparent that the Royal Society were still actively "monitoring" experiments on refractory materials after Dwight's death as Dr. Alexander Cumming FRS was sent to Georgia in North America in 1729/30 to assess the clay pits there and he negotiated a treaty with the Cherokees for supplies of their *unaker* clay—which Andrew Duche in 1732 was then able to use at Bow (Daniels, *The Origin and Development of Bow Porcelain, 1730–1747, Including the Participation of the Royal Society, Andrew Duche and the American Contribution,* 2007). The fact remains, however, that no porcelain was made at the Fulham Pottery after John Dwight's death in 1703, and we should seek a reason for this. Perhaps, John Haslem has summarised the situation nicely (Haslem, *The Old Derby China Factory,* 1876):

> In 1671 John Dwight had a patent for having by his own industry and at his own proper costs and charges invented and sett up at Fulham the mistery of Porcelain or China and Persian ware. Again in 1684 he protected the mistery of porcelain and other wares never before made in England or elsewhere. It would seem that the Doctor was in advance of the requirements of his day as his invention met with such poor encouragement that he burnt his receipts and implements in disgust and it is doubtful whether there is a specimen of his Porcelain extant. It is also said that he buried all his models, moulds and tools in some secret place on the premises at Fulham, observing that the production of such materials was expensive and unremunerating and he wished therefore to put it out of the power of his successors to perpetuate the business?

This may be a rather speculative sequel from John Haslem, but much of what he recounts in his work has been supported by later historical assessment (Reynolds 1862; Chaffers 1865): what is certainly true is that in many early porcelain manufactories there was a culture of the founder and driving force taking his secrets to the grave, Francisco Medici, Duke of Tuscany, at Florence and Louis Poterat at Rouen being just two of these. However, John Dwight must have made a significant number of finished and completed specimens for Hooke, Sloane and several other observers from the Royal Society to describe its ceramic body properties so accurately in terms of its potting, robustness to mechanical shock, translucency and heat resistance—the self-same characteristics for which the Chinese hard paste china then being imported into Europe was so highly esteemed. From the data reproduced in Table 6.1. It is seen that the Fulham stoneware has many characteristics of the later Coade stone that are nevertheless differentiated analytically, but it is rather higher in silica content but with similar medium-high alumina and alkali oxides, low calcia, low magnesia, low titania and low iron oxide content. It was perhaps not designed to be utilised for architectural purposes but must bear an acknowledgement for the manufacture of the first bust figures historically in artificial porcelain and stoneware of the bust of Lydia Dwight, John Dwight's daughter who died aged 6 in 1674, and also of Prince Rupert. The fine detail and moulding on these busts is especially noteworthy (Stanford 2020). The bust of Prince Rupert is especially interesting as Prince Rupert of the Rhine Palatinate (1619–1682) commanded the Royalist cavalry during the Civil War from 1642 and was a nephew of King Charles I. His bust was modelled by Edward Pearce (Pierce) (1635–1695), who died in 1695, and Prince Rupert died in 1682 and so Dwight may have modelled his bust somewhat earlier than this. Prince Rupert was a scientist and a founder member of the Royal Society, which would have brought him into personal contact with John Dwight and his experiments through both Robert Hooke and Robert Boyle. Prince Rupert was interested in the application of pigment on marble statues and stone and this too would have been of interest to John Dwight; the busts of Lydia Dwight and Prince Rupert were not commissioned by external clients but were retained as personal family heirlooms by John Dwight's family which means that there probably was a personal connection between their modelling and John Dwight—for his daughter Lydia that is taken, but Prince Rupert must have been a valued and respected scientific associate (he was renowned for his interest and support of scientific endeavours). A print of the bust of Prince Rupert in his regalia of the Order of the Garter is shown in Fig. 3.3, his full titles being Count Palatinate of the Rhine, Duke of Cumberland, Duke of Bavaria, and the Earl of Holderness.

3.3 Fired Clays

Although mud bricks have been known since 7000 BCE, where in the Middle East building bricks comprising mud and sometimes straw were shaped and dried in the sun, fired bricks which were formed at elevated temperatures appeared around 3500 BCE in the Mohenjo-Daro and other civilisations (Iqbal 2021). The Romans brought fired bricks to Britannia in 43 CE and these were composed of clay and sand composites that were fired in kilns at temperatures of around 600 °C. Terracottas also appeared around this time—specifically, a porous, clay- based fired unglazed ceramic which varied in colour from brown to orange and red in colour dependent upon the content of iron (III) oxide in the sand used for their manufacture. The higher the content of haematite in the sand raw material used then the deeper the red colour of the fired terracotta artefact. Glazed terracottas are usually termed earthenwares and these find more application in architecture and in household implements such as

Fig. 3.3 Print of the stoneware bust of Prince Rupert of the Palatinate, cavalry commander under King Charles I during the Civil War, Fulham Pottery of John Dwight. Formally dated to 1703 but since Prince Rupert died in 1693 and John Dwight in 1703, and its modeller Edward Pearce in 1695 it is probably best to consider it being completed before that date. Designed by John Spencer in the early 1690s, it shows the incredible detail in high resolution that high quality stoneware can take even at this early stage of its development. From the print in the New York Public Library, Miriam and Ira Wallach Division of Art Prints and Photographs, (Airy O. Charles, 1901); the original stoneware bust is in the British Museum (C.W. Reynolds Collection, 1871)

cooking vessels or storage jars for oil and wine as they are then impervious to water and liquids. Other terms have been adopted for glazed terracottas or earthenwares, such as faience, majolica and delftware, depending upon the type of glazing used, whether all the artefact was glazed (front and reverse) and its original place of manufacture. Earthenwares technically can be fired in a kiln operating up to 1000 °C but essentially they comprise only clay and sand as component raw materials. Perhaps the most famous terracotta artefacts known archaeologically are those of the so-called terracotta army, some 8000 life-sized figures of soldiers, cavalrymen, servants and horses which were constructed for the mausoleum of the Emperor Qin Shi Huang in 210 BCE at Xian in China (Quinn et al. 2017).

3.4 John Smeaton

A different line of development occurred in the late eighteenth century when John Smeaton FRS (1729–1792), known as the father of civil engineering, was commissioned to build the Eddystone Lighthouse in 1756, an account of which appeared in 1791, just a year before he died (Smeaton, *A Narrative of the Building and a Description of the Construction of the Eddystone Lighthouse with Stone to which is Subjoined an Appendix giving some Account of the Lighthouse on the Spurn Point built upon a Sand,* 1791). He was the first man to label himself a civil engineer and the forerunner of the Institute of Civil Engineers was the Smeatonian Society, which was founded in 1771. An integral part of Smeaton's plan for the design and construction of the Eddystone Lighthouse was its tapered shape and the development of a hydraulic cement for the foundations that would set under water (Nicholson, *Rock Lighthouses of Britain: The End of an Era,* 1995). The concept of a hydraulic cement that was resistant to water was not novel and historically can be traced back to the Greek lime-based *pozzolan* and then to the Romans in the third century BCE, who developed a sophisticated *harena fossicia, opus caementicum* and *pozzuolana* specifically for lining their aqueducts and construction of harbours. In Latin it was known as *pulvis puteolanis* and comprised a red volcanic ash from Pozzuoli near Naples and Santorini (giving it the alternative name of *Santorini earth*) and the Romans declared their recipe as two parts of *pozzuolana* which is mixed with one part of hydrated calcined lime and then mixed with ground travertine or marble to form an aggregate which did not involve sand. This material could cure under water and therefore was a cold-cured cementitious composite that did not require further firing. A large number of Roman buildings that used *pozzuolana* still survive today as testament to its reputation, including the Baths of Caracalla and the dome of the Pantheon in Rome built by The Emperor Hadrian in 125 CE, which has the largest unreinforced concrete dome from the Ancient World with a diameter of 142 feet and with a 27 foot oculus in the centre. In the port of Cosa, some 140 km north of Rome on the coast of Tuscany, founded in 273 BCE, the Romans made five piers of concrete which they poured directly underwater and these still survive today (McCann 1994; Jackson 2014). The volcanic ash used in *pozzuolana* contains a medium-high content of silica ranging from 50% to 65% and alumina as the other major component present around 30% (both similar to Coade stone in composition), with only a very minor content of iron (III) oxide and magnesia. Upon mixing with the hydrated lime it is converted into a calcium silicate and calcium aluminate. The recipe for *pozzuolana* cement was lost after the fall of the Roman Empire in 476 CE but was re-discovered some 1000 years later. The basis for later concretes was rather different in that although the hydrated lime in the form of Portland cement played a major role, and aggregates with limestone are still used, the other major component is now fine sand.

3.5 *Pozzolans* and Hydraulic Lime

It is appropriate here to distinguish between *hydraulic lime* and *hydrated lime*, as both terms have been applied historically and sometimes erroneously to concoctions of lime, clays and water used in ancient cements and mortars. The distinction between the two actually rests upon the method of setting and the chemistry involved: *hydraulic lime* (HL) is calcium oxide (quicklime) which forms calcium hydroxide (slaked lime) upon reaction with water only—setting by hydration — whereas *non-hydraulic lime* (NHL), also called *hydrated lime* or *lime putty*, sets and hardens by the absorption of carbon dioxide from the atmosphere and also requires moisture to effect this process (dry calcium oxide will not react with carbon dioxide). NHL is therefore not simply hydrated lime (Blezard 2003). The chemical formulae and reactions are given below:

1. The production of lime from limestone in the kiln:

$$CaCO_3 \rightarrow CaO + CO_2 \text{ at a temperature of} > 700\,^{\circ}\mathrm{C}$$

2. For hydraulic lime (HL):

$$\underset{\text{Quicklime}}{CaO} + H_2O \rightarrow \underset{\text{Slaked lime}}{Ca(OH)_2}$$

3. For non-hydraulic lime (NHL):

$$\underset{\text{Quicklime}}{CaO} + H_2O + CO_2 \rightarrow \underset{\text{Slaked lime}}{Ca(OH)_2} + \underset{\text{Calcite}}{CaCO_3}$$

The HL mixtures often had additions of clays, sand and *pozzolans* and are referred to in Vitruvius as *Roman cement (opus caementicum)* (Marcus Vitruvius Pollo, Vitruvius, *De Architectura*, transl. by Morris Hicky Morgan as *The Ten Books on Architecture*, 1914). If volcanic ash was not immediately available from Vesuvius, Etna or Santorini then an effective substitute was deemed to be finely ground terracotta or pottery and sand, a composite which was then called *coccoiopesto*.

Historically, three types of HL were recognised, dependent upon their empirical setting times, all containing *pozzolan* and clay added to sand and quicklime (Massazza 2003): Feeble HL containing <10% clay, taking one week to set upon the addition of water; Moderate HL, containing 11–20% clay, taking few days to set upon the addition of water; Higher HL, containing 21–30% clay, taking within one day to set upon the addition of water. All had the property of setting under water (thereby excluding the ingress of carbon dioxide) unlike their non-hydraulic lime (NHL) versions which required the advent of both carbon dioxide and water to effect a set—and were therefore normally used for building mortars and for the lime putty used in Roman frescoes and wall plasters and which needed excess water to render them malleable and workable after application. All these lime preparations gained strength with time and curing (Xu et al. 2021). The composition of the pozzolanic cement mixtures was variable and John Smeaton in his construction of the Eddystone Lighthouse in 1756 adopted a recipe comprising 50% of quicklime with 50% of a pozzolanic /clay/sand mixture containing 11% clay: the imported *pozzolan* he purchased from a merchant in Plymouth for use on the Eddystone Lighthouse—several courses of stone were permanently laid under the sea level and others laid on top could be subjected to constant wetting from seawater and spray. Smeaton's mixture as used for the Eddystone Lighthouse approximates to the Moderate HL classification of hydraulic and pozzolanic lime described above.

When additional components were incorporated as raw materials into the mixture to be fired the descriptors of the product can be confusing. It is generally accepted that the manufacture of porcelain was discovered in China during the Tang Dynasty sometime after 600 CE and after some centuries of experimentation it reached a perfected form around 1200 CE. In common with many porcelain manufacturing enterprises that appeared later in history, the successful Chinese porcelain syntheses grew from practical experience in the manufacture of pottery, faience and celadons and the application of glazes to terracottas and earthenwares. Thereby, the skills in firing clays and composites at high temperatures in kilns were acquired empirically by a local workforce long before the advent of chemical analyses and scientific information about the behaviour of materials at elevated temperatures was available to assist in the production of ceramics.

References

J.M. Blashfield, *A Catalogue of Seven Hundred Articles Made in Patent Terra Cotta, and Red and Cane-Coloured Pottery* (Henry Johnson, Stamford, 1882)

R.G. Blezard, Chapter 1: the history of calcareous cements, in *Lea's Chemistry of Cement and Concrete*, ed. by P.C. Hewlett, 4th edn., (Elsevier/Butterworth Heinemann, Oxford, 2003), pp. 1–23

W. Chaffers, The earliest porcelain manufacture in England and imitation of Cologne Ware in the 17th century. Art. J. (June 1865 issue), 179–181 (1865)

Sir A.H. Church, *English Porcelain: A Handbook to the China Made in England During the 18th Century as Ilustrated by Specimens Chiefly in the National Collection* (A South Kensington Museum Handbook, Chapman & Hall Ltd., London, 1885 and 1894)

P. Daniels, *The Origin and Development of Bow Porcelain, 1730–1747, Including the Participation of the Royal Society, Andrew Duche and the American Contribution* (Resurgat Publishers, Oxford, 2007)
H. Eccles, B. Rackham, *Analysed Specimens of English Porcelain in the Victoria and Albert Museum Collection* (South Kensington Museum Press, London, 1922)
H.G.M. Edwards, *Porcelain Analysis and Its Role in in the Forensic Attribution of Ceramic Specimens* (Springer-Nature, Dordrecht, 2022)
A. Fontaine, *Der Englisch Coade -Stein – die Villeroy & Boch "Terakotta", mit Excursen zu Martin Klein nach Weimar und Friedrich Schinkel nach Berlin* (Books-on-Demand Publishing GmbH, Norderstedt, 2019)
C. Green, *John Dwight's Fulham Pottery: Excavations, 1971–1979*, Archaeological Report 6 (English Heritage, London, 1999)
D. Haselgrove, J. Murray, John Dwight's Fulham pottery, 1672–1978: a collection of documentary sources, J. Ceram. Hist. 11, Stoke-on-Trent City Museum (1979)
J. Haslem, *The Old Derby China Factory* (George Bell & Sons, London, 1876)
D.A. Holdridge, Ball clays and their properties. Trans. Br. Ceram. Soc. **55**, 369–440 (1956)
J. Horne, *John Dwight "The Master Potter" of Fulham, 1672–1703: Catalogue of an Exhibition at Kensington Church Street* (John Horne Publishing, London, 1992)
S. Inwood, *The Man Who Knew Too Much: The Strange and Inventive Life of Robert Hooke, 1635–1703* (Macmillan, London, 2002)
S. Iqbal, Energy efficiency in architecture design in Mohenjo Daro, in *Cities' Vocabularies: The Influences and Formations*, ed. by N. Mohareb, A. Versacci, Y. Mahgroub, S. Maruthaveeran, F. Albert, (Springer Nature, Cham, 2021), pp. 203–208
M.J. Jackson, Sea-water concretes and their material characteristics, in *Building for Eternity: The History and Technology of Roman Concrete Engineering in the Sea*, ed. by C.J. Brandon, R.L. Hohlfelder, M.D. Jackman, J.P. Oleson, (Oxbow Publishers, Oxford, 2014), pp. 141–187
D.A. Long, *"At the Sign of Atlas": The Life and Work of Joseph Moxon, A Restoration Polymath* (Shaun Tyas, Donington, 2013)
Y. Maniatis, M.S. Tite, Technological examination of Neolithic -Bronze age pottery from Central and South Eastern Europe and from the Near East. J. Archaeol. Sci. **8**, 50–76 (1881)
F. Massazza, Chapter 10: pozzolana and pozzolanic cements, in *Lea's Chemistry of Cement and Concrete*, ed. by P.C. Hewlett, 4th edn., (Elsevier/Butterworth Heinemann, Oxford, 2003), pp. 471–635
A. McCann, The Roman Port of Cosa (273 BC). *Scientific American: Ancient Cities*, 92–99 (1994)
C. Nicholson, *Rock Lighthouses of Britain: The End of an Era* (Whittles Publishing, Caithness, 1995), pp. 33–43
W. Page, *A History of the County of Middlesex : General – Ashford, East Bedfont with Hatton, Feltham, Hampton with Hampton Wick, Hanworth, Laleham, Littleton*, vol 2 (Victoria County History, London, 1911), pp. 142–146
A. Peart, The Fulham Pottery, 1932–1965: a return to artistic production. J. Decor. Arts Soc., 1850-The Present **32**, 101–125 (2009)
P.S. Quinn, S. Zhang, Y. Xia, X. Li, Building the Terracotta army: ceramic craft technology and organisation of production at Qin Shihuang's Mausoleum complex. Antiquity **91**, 966–979 (2017)
W.R.H. Ramsay, E.G. Ramsay, *The Evolution and Compositional Development of English Porcelains from the 16th Century to Lund's Bristol c. 1750 and Worcester c. 1752- The Golden Chain* (Invercargill Press, New Zealand, 2017)
C.W. Reynolds, Fulham pottery. Art J. (October 1862 issue), 204 (1862)
J. Smeaton, *A Narrative of the Building and a Description of the Construction of the Eddystone Lighthouse with Stone to Which is Subjoined an Appendix giving some Account of the Lighthouse on the Spurn Point built upon a Sand* (Longman, Hurst, Rees, Orme and Brown, Paternoster Row and H. Hughs, Pall Mall, London, 1791)
M.L. Solon, *Art of the Old English Potter* (Bemrose & Son, London, 1883), pp. 32–35

M. Spataro, M. Bimson, N. Meeks, Part3: a scientific investigation of three pieces of porcelain presumed seventeenth century, from Burghley House. Trans. Engl. Ceram. Circle **20**, 187–194 (2008)

M. Spataro, N. Meeks, M. Bimson, A. Dawson, J. Ambers, Early Porcelain in 17th century England: non-destructive examination of two jars from Burghley House. Br. Mus. Res. Bull. **3**, 37–48 (2009)

C. Stanford, *The mistery of Cologne or Stonewares: were John Dwight and Robert Hooke the pioneers in the invention of fired artificial stone in architecture?* (Georgian J., XXVIII, 2020), pp. 1–20

M.S. Tite, I.C. Freestone, M. Bimson, A technological study of Chinese porcelains of the Yuan Dynasty. *Archaeometry* **26**, 139–154 (1984)

M.S. Tite, M. Bimson, I.C. Freestone, A technological study of Fulham Stoneware. In: Proceedings of the 24th Archaeology Symposium (Washington DC, USA, 1986), pp. 95–104

M.V.P. Vitruvius, De Architectura, The Ten Books on Architecture, 15 BCE, Translated by Morris Hicky Morgan (Harvard University Press/Oxford University Press, Cambridge, MA/Oxford, 1914)

K. Xu, A.S. Tremsin, J. Li, D.M. Ushizima, C.A. Davy, A. Bouterf, Y.T. Su, M. Marroccoli, A.M. Mauro, M. Osanna, A. Telesca, and P.J.M. Monteiro, "Microstructure and water absorption of ancient concrete from Pompeii: an integrated synchrotron microtomography and neutron radiography characterization", Cem. Concrete Res., 139, 106282, 2021.

Chapter 4
Historical Myths and Anomalies Associated with Coade Stone

Abstract Here, the historical myths and legends associated with Coade stone, Eleanor Coade, and her eponymous artificial stone are discussed which have been perpetuated incorrectly or accounts been misinterpreted in texts and articles hitherto. These include the secrecy of the recipe, its unknown composition, Eleanor Coade's patent, the precise location of her manufactory, the role of her mother (also Eleanor Coade), the constancy of the composition of Coade stone, exaggerated claims of its survivability in harsh environments and the proper identification of genuine Coade stone artefacts are examined.

Keywords Coade stone patent · Mrs. Eleanor Coade · Coade stone recipe and formulation · Manufactory location · Festival of Britain · Westminster bridge · Narrow Wall · King's arms stairs

The upsurge in desirability for Coade stone in the eighteenth century and its ready adoption by the leading architects, landscape gardeners and practitioners of the time has to some extent obscured several potential anomalies that have arisen about its invention and production and has created some myths and legends surrounding Eleanor Coade and her artificial stone material that need to be evaluated in a modern forensic holistic assessment of the phenomenon. Several myths have arisen from the statements made by historians of the great houses where Coade stone is located and of the families who commissioned these artefacts, and a listing of the relevant Coade stone literature is given in Appendix III. There is no doubt that the appearance of Coade stone artefacts towards the closing decades of the eighteenth century and their wide adoption by the most influential designers, landscape gardeners and architects of the day did much to create a diverse history associated with Eleanor Coade and her artefacts that were made from this wondrous new material and did much to generate ideas and thoughts which actually have no practical basis in fact or reality. Some of these myths and legends surrounding Eleanor Coade and her eponymous artefacts will now be discussed and confirmed or corrected below. Whereas several of these allegations can be dismissed fairly simply as a complete fabrication, others do need a fuller investigation to establish the worthiness of their consideration on a factual forensic basis and to set straight the historical record.

H. G. M. Edwards, C. J. Brooke, *Coade Stone*, Cultural Heritage Science,
https://doi.org/10.1007/978-3-031-71432-0_4

4.1 Eleanor Coade Invented Artificial Stone

This statement is commonly made in the literature and it is completely false. Artificial stone preparations had been known for hundreds of years prior to Eleanor Coade (*viz.* Roman cement). An historical statement has been made dating from the reign of King Henry VIII that the overseer of repairs to one of his palaces used materials made from an artificial stone, some two hundred and fifty years before Eleanor Coade appeared on the scene, although tantalisingly no information is given as to the nature of this material, which was probably a variant of a Roman cement but using local raw materials. Eleanor Coade certainly did not invent fired ceramic stone or even the phrase "*lithodipyra*" (twice-fired stone), which had been used by Daniel Pincot two years before Eleanor bought his artificial stone business (Pincot 1770). Yet, the first recorded and dated impressed mark on a Coade stone artefact was COADE LAMBETH LITHODIPYRA in 1787, which could well have helped to sustain the origin of this particular myth! Other related claims centre upon the statement that still appears in the modern literature that Eleanor Coade was responsible for the formulaic recipe of her eponymous stone, without her having had any experience at all hitherto of ceramics synthesis. Such statements as "*Mrs Coade devised a formula to mass produce architectural embellishments and statuary of the highest quality*" seem quite incontrovertible superficially with what we currently know about Coade stone but actually upon inspection do not have any foundation in documents regarding her supposed invention of the formulaic recipe she used at first hand . . . that was most certainly Daniel Pincot's prior role. When one examines the empirical trials and difficulties that were experienced by the early manufacturers of synthetic porcelain in achieving their competitive products, which generally were only accomplished after several setbacks, it does indeed seem incongruous to assume that Eleanor Coade, who came from a textile sales background with no previous experience at all of the practicalities and intricacies of high-temperature ceramics manufacture with its many associated failures in formulation and processing, could take on a failing synthetic artificial stone business, modify chemically the formulation of its composites strategically and immediately secure the prime business position in a competitive market—and moreover, all this was done remarkably rapidly after her acquisition of Pincot's failing business in 1769, as she was supplying clients with manufactured pieces in 1771!

4.2 The Secret Recipe for Coade Stone Has Been Lost

A dramatic statement that again is not true; whereas there is little evidence that the detailed recipe and formulation for Coade stone was published during Eleanor Coade's lifetime, apart from the revelation by David Laing of its raw materials in 1821, it was certainly published in full by her distant relative Charles Fowler in 1850 after closure of the Coade & Co. manufactory (Fowler 1850), but prior to this it was known and indeed used by her successors Mark Blanchard and John Marriott Blashfield in the 1840s, who both set up ceramic artificial stone businesses using her designs, recipe and ideas, and indeed even her original moulds for the re-creation of her products. It seems to be a fabrication that the Pincot/Coade artificial stone recipe was ever a "secret" although it was never announced as such publicly. Another statement that appears frequently in the literature is that modern chemical analysis has now resulted in the discovery of this long lost recipe for Coade stone (Kelly, *Mrs Coade's Stone*, 1990): this again is untrue, and although the first analyses made of a single Coade stone specimen from an unspecified location in Belmont House were undertaken by Professor Ian Freestone in 1991, the recipe had already been known for 140 years when he published the results of his elemental oxide analyses, which were found to be entirely consistent with the published recipe as he correctly stated in his paper (Freestone 1991), referencing therein the prior publication source of that recipe by Fowler (1850). Even recently, it has been alleged (Foyle, *Financial Times*, 2015) that some extensive research over 35 years was necessary to determine the recipe—the article entitled enticingly "How a sculptor cracked the lost mixture of Eleanor Coade's stone". As an aside, it is generally not a simple matter to deduce a precise recipe composition given in component raw materials backwards from the analytical determinations of the elemental oxides because of the high-temperature chemical reactions undergone in the kiln, which alter the mineral compositions of the raw materials beyond recognition (Edwards, *From Porcelain to Silica Bricks: the Extreme Ceramics of William Weston Young, 1776–1847*, 2019), although analysts can apply inductive reasoning to confirm whether or not their data are consistent with a particular recipe that has been so formulated. This demands a specific knowledge of the kiln temperature achieved for the particular ceramic synthesis and of the chemical reactions that have taken place in the achievement of that temperature (Edwards, *Porcelain Analysis and Its Role in the Forensic Attribution of Ceramic Specimens*, 2021) and this is certainly not a trivial task to inductively deduce the raw materials used in a recipe from the analysis of a ceramic body.

4.3 The Coade and Sealy Works Site Is Buried Beneath the Royal Festival Hall

This statement is actually only partially correct: strictly, when the original Coade and Sealy works site at Westminster Bridge was cleared for the Festival of Britain in 1951, the remains of the artificial stone works actually became buried beneath the Jubilee Gardens, South Bank, north of Westminster Bridge, and not immediately under the Royal Festival Hall. The Coade and Sealy site lay beneath the Dome of Discovery and the Skylon on the Festival of Britain site, both of which were soon adopted as iconic symbols of the Festival of Britain. The Royal Festival Hall already existed when the Festival of Britain opened in 1951 and was built as part of the

Festival of Britain site in 1949, the same year that the Coade manufactory site was developed for this purpose. The Jubilee Gardens, on the other hand, were later opened on the old Festival of Britain site to commemorate the Silver Jubilee (25 years on from June 1952) of the accession of Her Majesty Queen Elizabeth II in 1977, some 28 years later. The large grindstone that was used by Eleanor Coade and her successors to produce the necessary very fine frit from the raw materials for the artificial stone paste destined for the moulds has been preserved and is now mounted beneath an abutment pier of Westminster Bridge, near to the old manufactory site (Roberts & Godfrey, *Survey of London: Lambeth: South Bank and Vauxhall*, 1951).

4.4 Eleanor Coade Inherited the Coade Artificial Stone Business from Her Mother?

This statement grew from the assumption by earlier historians that all references in early documentation to a "Mrs. Coade" must have referred to Eleanor's mother, Eleanor Sr, as Eleanor her daughter had never married. However, "*Mrs*" was in fact a courtesy title accorded to Georgian businesswomen regardless of their marital status. The only definitive documents relating to Eleanor Sr and her role in the business are two invoices for the Poor Rates for Eleanor Jr. and her Mother at their home in Camberwell, where Eleanor Sr was referred to clearly therein as Mrs. Coade, Widow. It has also been suggested from this assumption that Eleanor Sr bought the artificial stone business from Daniel Pincot in 1769 and ran this until she died aged 88 in 1796, when it then passed to her daughter, Eleanor Jr., who then took her cousin John Sealy into partnership in the business as Coade & Sealy. There is a chronological discrepancy here because other documentation refers to Eleanor purchasing the business directly from Daniel Pincot in 1769 and within the year dismissing him for dishonesty, then taking on John Sealy the self-same year, i.e. 1771, into a partnership. Coade & Sealy then became the official business name although they did not launch their first Catalogue of artefacts until 1784. The source of the money put up to purchase the business from Pincot, namely £4200, has also been investigated and it is believed that Eleanor Jr. provided the sum, possibly with the aid of some input from Eleanor Sr, but it should be remembered that George Coade, Eleanor's father, had been declared bankrupt for the second time a year or so earlier and that a previous bankruptcy had occasioned the relocation of Eleanor and her sister Elizabeth to another residence elsewhere to avoid being embroiled in the family financial liabilities, which surely would have impinged upon Eleanor Coade Sr as well, as she was then resident in the family home. Clearly, Eleanor Sr would not have had the ability on her own account to finance the purchase of another business at that time. It is also quite possible that Eleanor Jr. would have sought financial aid from the Nonconformist community (Dissenters) of which she was a member in Camberwell, along with Daniel Pincot, to assist in the purchase of

Pincot's business as the organisation was noted for supporting its members financially in such ventures.

The following statement was taken from an historical website which was accurate in its description of the surviving Coade stone artefacts but otherwise was patently very incorrect and creates some confusion when details are being compared with the real situation:

> Mrs Coade was a resident of Lyme Regis and lived at Belmont House. Despite the presence of Coade stone at several prominent sites, after Mrs Coade's death in 1821 her daughter and relatives who inherited the firm did not do as well and the firm went bankrupt in 1833.

The chronology is completely wrong here: firstly, the assumption is that Mrs. Coade was Eleanor Sr, who was the businesswoman who purchased and ran the firm until her death in 1821 (at the age of 113, apparently, as she was born in 1708!). Eleanor Jr. would have inherited the firm from her mother (who had died 25 years before) then in the same year she also died at the age of 88. Other relatives did not inherit the firm afterwards as it was bequeathed to charities .... And the business went into bankruptcy in 1835 not because of inept business behaviour but rather the occurrence of bad debts and the increased competition with the much cheaper cementitious cold-cured stone artefacts that were then coming onto the market. Finally, the assumption that Eleanor Sr lived at Lyme Regis in Belmont House cannot be sustained and this is total fabrication—Belmont House only passed to Eleanor Coade (Jr!) in 1784, when Eleanor Coade Sr would already have been 76 years of age. Eleanor regarded Belmont House, or Bunter's Castle as it was then known to her at that time, as her holiday home, and it became her own "seaside pavilion", as she remarked.

4.5 Did Coade Stone Production Generate Waste?

Firstly, being a ceramic material, Coade stone was a primarily a moist paste of the raw materials that was moulded and fired at high temperatures in the range between 1100 and 1200 °C. Early analyses from the British Museum Research Laboratory in 1984 of relics excavated from the Coade stone manufactory site prior to the location there of the Festival of Britain in 1951 have established the composition and have determined that the kiln firing temperature of Coade stone took place between 1100 and 1150 °C. To any researcher with a background in porcelain synthesis of the manufactories of the eighteenth century this immediately suggests that process control, especially kiln temperature control and maintenance, would have been paramount for a successful thermal outcome of the fired artefacts to be achieved but, nevertheless, there would still be a significant wastage derived from mis-fired items, which then had to be broken up and consigned to a waste tip, perhaps sold off or maybe used internally in the manufactory as "grog" (Edwards, *Porcelain Analysis and Its Role in the Forensic Attribution of Ceramic Specimens*, 2021). Scientific research into early porcelain synthesis has established the critical need to achieve thermally the eutectic region of the ternary or quaternary oxide diagrams, which relate the interaction of the major components of the porcelain paste with each other chemically at elevated temperatures: generally, the eutectic, or *liquidus* region, is defined by a rather narrow range of temperatures wherein the specified mineral species can exist together in a thermally stable state of vitrification and outside these temperature limits the fired ceramic artefacts are found to be deficient through overheating (causing a sagging or a deformation of their shape) or underheating, resulting in an incomplete vitrification process and domains in the ceramic body of incompletely converted crystallites (thereby causing cracking, fissures and artefact textural problems—the so-called "firing cracks"). All porcelain manufactories were bedevilled by mis-fired batches from their kilns which resulted in a significant wastage of product. Technically, a vertical thermal range or temperature ramp operated within the kilns, with the hotter parts being uppermost and the lower parts being cooler and a horizontal temperature difference was also manifest between the thermal brick-lined walls and the centre of the kiln. The use of a good coal as fuel to achieve the high temperatures necessary in the kiln was critical to achieve the firing quality of the artefacts enclosed therein: William Billingsley at the Nantgarw China Works was fortunate to have the best coal available from the Rhondda Valley, namely Welsh steam coal or anthracite, which possessed the largest thermal output and lowest ash for its weight, and thereby he was able to maintain his kiln at a temperature in excess of 1300 °C for several days at a time, which was at the upper limit for porcelain synthesis at that time. The integral placement of the incipient ceramic artefacts awaiting firing in a kiln was a specialised task and experienced kiln managers were highly respected and valued for their empirical skills as the proprietors of the porcelain manufactories were well aware of the economic production losses that resulted from the mis-firing of the artefacts due to a kiln processing failure. Eleanor Coade is recorded as having had several contretemps with her unnamed kiln manager(s), who were apparently found to be in the habit of leaving their posts during an extended firing sequence overnight and adjourning to a local hostelry near the manufactory, so she clearly appreciated the problems that could arise from unsupervised kiln process mis-firings. The presence of a trustworthy and experienced kiln manager was essential for ceramics production as there were no thermometers available at that time which could register the high temperature of the kiln or furnace; kiln managers therefore estimated the temperature of their furnaces by eye and through the redness in colour of the thermal brick lining of the kilns, although the idea of using ceramic saggars and cones which softened over several narrow empirical temperature ranges were advocated for an aid to this visual kiln temperature assessment. It is believed that a primitive type of pyrometer was being developed at this time by Josiah Wedgwood at his ceramic manufactory in Etruria, Staffordshire, but this certainly would not have been generally available and would have required special expertise in its use and adoption for furnaces. It is interesting too, that contrary to the historical practice adopted by contemporary porcelain manufactories, Eleanor Coade did not reveal the names of her personnel involved in the manufacture of her Coade stone: of course, we know several of the names of her sculptors, artists, architects employed and designers of her statuary but the names of her artisan mould makers, finishers (settlers), kiln foremen, composite paste makers and kiln managers are not known.

Porcelain manufactories all had waste pits for the dumping of overfired and underfired ceramic artefacts from kiln mis-firings and these have provided a rich source of materials for archaeological excavation and subsequent analytical studies into the recipes and changes that were made during the production lifetimes of the manufactories concerned. It is very surprising, therefore, that no historical account of Eleanor Coade's business operation makes reference to similar waste pits and the materials therein, or to kiln losses and to the destination of substandard artefacts emerging from the firing processes. Compared with porcelain manufactories such as those at Sevres, Derby, Nantgarw and Swansea, where kiln losses were typically between 25% and 90% of the ceramics production, an apparently implied zero % age kiln loss for artificial Coade stone ceramic analogues would be as a result incredible and just not attainable. Even with the several centuries of first-hand experience possessed by the Chinese ceramic manufacturers at their Imperial porcelain site at Jingdezhen, European visitors to the site in the eighteenth century commented upon the very large waste tips seen there, which some observers thought possibly represented as much as 50% of their fired porcelains. This must have been a huge amount of porcelain waste as the famous dragon kilns at Jingdezhen could accept up to 25,000 items of hard paste porcelain artefacts for firing in a single charge, so a single kiln failure represented a very large amount of waste material. William Duesbury at the Derby China Works used to purchase "grog" waste from other manufactories in ton quantities (Anderson 2000) as a supplementary frit addition to his soft paste porcelain pastes prior to their moulding and final firing and perhaps this could offer an plausible explanation for the omission of the mention of a waste pit at the Lambeth manufactory site: quite simply, Eleanor Coade could perhaps have recycled many of her damaged and imperfect items as a grog component for the next batch of artificial stone artefacts destined for production from the Lambeth manufactory? This would certainly explain the absence of a waste pit as the grog from a previous unsuccessful firing sequence would have been largely consumed in the production of the ceramic paste for the next firing sequence, perhaps a week or so later, and this may well have involved the use of her large granite grindstone for this purpose, which has survived demolition of the site and has now been preserved as a relic on the factory site near Westminster Bridge (H. Roberts and W.H. Godfrey, *Survey of London: Lambeth: South Bank and Vauxhall*, 1951). Another major advantage of using one's own grog in this way is that its composition would be well—known and, being unglazed, would not suffer from contamination by the heavy metal tin or lead metal oxides, which would perhaps have negated such a usage of grog supplied from the contemporary porcelain manufactories which may additionally have used an unspecified component of grog from several other contemporary ceramics sources.

John Ruch (Ruch, *Regency Coade: A Survey of the Coade Record Books, 1813–1821*, 1968) has researched the handwritten documents preserved in the Public Records Office, London, relating to the commissions undertaken by the Coade manufactory between 1813 and 1821 under the stewardship of William Croggon:

these detail the correspondence between the manufactory and clients from the order through construction to the completion and accounts, along with Croggon's work notes, in three volumes, including a list of noble and titled clients (i.e. nobility, knights and baronets). This provides a valuable basis for the determination of the original commissions of Coade stone artefacts, especially in the later years of the manufactory operation during the second decade of the nineteenth century. Several interesting nuances can be discerned which have generally not been revealed in previous historical accounts of Coade stone, for example:

- In February 1815, Sir Godfrey Webster (1789–1836), 5th Baronet of Battle Abbey and Copthall and MP for Sussex, ordered a coat-of-arms from the Coade manufactory: William Croggon requested a copy of the arms for this order when placed in the August of that year, but none was received so Croggon sent a representative to make a drawing of the coat-of-arms on the carriage of the client . Eventually the order was completed in the December and delivered to Sir Godfrey but the motto on the artefact was found to be faulty when delivered and inspected in February 1816. Sir Godfrey's motto was *Fides et Justitia,* meaning *Faith and Justice*, which seems straightforward enough, but presumably this error resulted in a dispute over settlement of the account (Fairbairn, *Crests of the Families of Great Britain and Ireland and Their Kindred in Other Lands,* Volumes I and II, 1905). This is the reason for the use of heraldic bookplates in armorial reconstruction and enamelling on porcelain for their correct representation, as has been highlighted by Denyer et al. (*Armorial Porcelains; The Genesis*, 2024).
- Several Coade stone figures and statues were commissioned for the Britannia Monument erected in Great Yarmouth in 1819 to commemorate Admiral Lord Nelson and his victory at the Battle of Trafalgar in 1805, see Chap. 7 and Fig. 7.58. It is recorded that the Coade stone artefacts on the top of this column, namely four caryatids supporting a figure of Britannia, had to be replaced unusually by cementitious replicas because of marine air erosion in 1896. William Croggon notes in his work book that one of the caryatids had developed a crack during the kiln firing but he did not regard this a serious occurrence and he despatched the artefact to its destination anyway but took a special precaution to carefully pack it during carriage. This might be a reasonable explanation for the deterioration of these particular Coade stone artefacts as Croggon also records that he had noticed that overfiring made the stone more durable but of a deeper colour whereas excessive moisture retention in the clay during its preparation caused a deterioration in its quality. It seems then that a variability in Coade stone product quality was known and appreciated by the proprietors at the time.
- Although the arrangements were put in place in 1798 for the establishment of The Coade Gallery and Exhibition in Artificial Stone on the site of the manufactory, which opened in 1799, where the prospective clients could view a range of statues and architectural exhibits and see a brochure which comprised a 12-page introduction followed by 24 pages describing the objects on show, it seems to be rather incorrect to describe this pamphlet as the "Coade & Sealy Catalogue of 1799" as have other historians (Esdaile 1940).

4.6 Were Coade Stone Artefacts Always Produced to the Same Standard?

The issue of the variability in standard of the quality of the Coade stone ceramic product is a much more difficult one to assess as, again, little comment has been made historically about this aspect, although there is a tacit assumption made that all artefacts emerging from the manufactory were of a similarly high quality. It is certainly true that in all other ceramic high-temperature fired artefacts, the quality did vary with the batch processing, within the batch and between successive batches, and this was dependent upon several factors such as the position of the artefact in the kiln which dictated its operational firing temperature, the effectiveness of the kiln thermal maintenance schedules during the firing process, the kiln temperatures achieved at each firing and small deviations made in recipe formulation which may have occurred through a change of specific raw material components, their alternative sourcing and deterioration of the raw material upon storage. The implication in the literature surveys of artificial Coade stone is that all Coade stone artefacts were consistently of the same high quality, but surely this cannot be accepted as being correct now without checking, given the variables that necessarily applied: this now begs the further question—what happened to those fired artefacts that were perhaps deemed to be substandard in quality at the manufactory and were they then destroyed or offered for sale as "seconds", at perhaps a cheaper price regardless of their perceived substandard quality, as happened in their contemporary analogous porcelain manufactories? For the Chinese Imperial porcelain manufactory at Jingdezhen in the late eighteenth century, the Emperor Kangxi issued an edict that his own porcelain commissions for his palace in Peking must be of the highest quality uniformly and that lower quality porcelains (which he rated as second and third grade) from the same kiln firings should be destroyed: however, the latter instruction was not always carried out to the letter by the kiln managers at the manufactory and an illicit trade in substandard artefacts was prevalent for subsequent purchase at reduced prices by predominantly local clients, as can be witnessed by their survival into modern times. There is little forensic analytical evidence to enlarge upon this hypothesis for artificial stone artefacts but several personal comments made to the authors by a modern restorer of several hundred Coade stone artefacts is that there is indeed a perceived variability in their quality, which has been reflected in their subsequent resistance to their sustained environmental damage, and also perhaps surprisingly that there could well be an indication of a potential variability of their composition too (Philip Thomasson, personal communication). This is an aspect that could be investigated by modern analytical techniques to provide information on this aspect of Coade artificial stone artefacts but would necessarily require the sampling of a large number of specimens, which may not be possible practically. In the meantime, the assertion of historians through the ages that Coade stone was impervious to environmental damage and that the only damage perceived to artefacts has been the result of demolition, vandalism or acts of war should be treated *cum grano salis*: however, the susceptibility of Coade stone artefacts to mechanical shock and rough handling will be evident in several of the Case Studies reviewed here in Chap. 7. There could be possibly other factors which would impact upon their survival capability and which could be independent of local environmental stresses and can be traced back to perhaps minor synthetic variations in their manufacture such as those that have been postulated above.

4.7 Was There a Variability in Coade Stone Composition During the Lifetime of the Manufactory?

The comment in the above section by an esteemed professional restorer and modern manufacturer of artificial stone for the repair of damaged Coade stone artefacts reinforces the idea that has been mooted historically that Eleanor Coade may have amended her recipe slightly during the years of operation of the manufactory and this is surely not in itself surprising although it does not lend favour to the suggestion that she immediately changed the formulation from that being operated by Daniel Pincot when she took over the manufactory from him in 1769. Various comments that have been made historically, such as "*Eleanor Coade varied the composition of her stone according to the size of the piece being fired*" are probably drawing upon known exemplars in the manufacture of analogues in the hard and soft paste porcelain industries that were operating at that time, whose proprietors were constantly trying to improve their products against those of their competitors. However, these changes, although maintaining a necessary secrecy outside the manufactory, were normally logged in daybooks kept by the proprietors themselves in the manufactory; this was an absolute necessity for proprietors to keep a check on their sequential alterations made to their body pastes or glazes without relying on memory . For example, Lewis Dillwyn FRS, of the Swansea China Works, kept a meticulous record of his experimental scientific trials with his own variable paste compositions undertaken with Samuel Walker, his kiln manager, over an 18 months period from 1815 to mid-1817, yet these were still written in his hand in his personal day book and also in code to protect the information from their involuntary access to competitors! It is inconceivable that Eleanor Coade in partnership first with Daniel Pincot, then John Sealy and finally with William Croggon, would not have recorded any vital raw materials data and changes similarly incurred in the recipe somewhere in a day book if such compositional changes were envisaged, undertaken and employed at several stages during the firm's 70 years of production history. What is more likely, perhaps, is that a change in geographical sourcing of her raw materials may have been necessitated occasionally during the production period which would have had a potential impact upon the quality of her fired ceramics and that this should also have been noted somewhere. It is, however, unlikely that a change in sourcing for the Coade stone raw materials would have had the same calamitous result that it did upon the Limehouse porcelain manufactory in 1747 when supplies of their essential raw material steatite, a magnesium silicate, from the Kynance Cove mine in

Cornwall, became worked out and dried up suddenly: several months later, in the following year, the Limehouse manufactory was forced to close down permanently as the product porcelain quality using an unsatisfactory replacement material had then become unviable against its competition from elsewhere.

In an earlier publication, Edwards (Edwards, *Porcelain Analysis and Its Role in the Forensic Attribution of Ceramic Specimens,* 2021) considered the effects of errors in the accuracy and precision of weighing the raw materials at the manufactory had upon the characteristic composition of the prototype paste for soft paste porcelains that was to undergo batch moulding and firing in the kilns. A factor that had not previously been considered and evaluated scientifically was the effect of moisture retention by the raw materials, which in most cases had been transported over relatively large distances in open waggons that were exposed to the prevailing weather for relatively long times, followed perhaps by storage under inadequate conditions at the manufactory site. It was estimated that an improper protection of the raw materials against the ingress of moisture and subsequent inaccuracies in weighing batches for making into the ceramic pastes could account for up to a 15% cumulative error in their quantitative recipe preparation. This is not insignificant and could well result in an unregulated batch variation that would be reflected in a real and measurable compositional change of the artificial stone paste even when none was even intended or appreciated by the manufacturer. On the other hand, it could be that Eleanor Coade subscribed to an increased vigilance and supervision in the quality control of her raw materials, as was manifest by some of her contemporary porcelain colleagues, which certainly would have obviated many detrimental factors which otherwise might have contributed to the imposition of such adverse and unexpected compositional changes in her fired product. This would be retrospectively well within her proprietorial purview and acumen in setting up her business operations and she could have had much open discourse with colleagues in the ceramics industry about this aspect of manufacturing which was well known to china manufacturers, even if they did not have at that time the scientific equipment currently available to quantifiably measure the moisture content of the minerals or the inadvertent acquisition of organic contaminants that had accumulated in their raw materials. A raw material which was particularly susceptible to atmospheric effects was quicklime, which reacts with both water and carbon dioxide to change its chemical integrity irreversibly at ambient temperatures: hence, porcelain and other ceramics manufacturers who used quicklime as a raw material fluxing agent in their paste had to recognise the possibility of the fundamental variability of their source material occurring. Eleanor Coade did not specifically use quicklime as a flux in her recipe formulation but relied upon the small calcia component present in her ground soda lime glass to provide a high temperature flux ... but this soon became a critical factor for the later cementititous artificial stone manufacturers such as Austin and Seeley, where a major component of their recipe was quicklime and any deterioration in its quality was to have a significant effect upon the resultant product. It is known that Austin and Seeley's ambient temperature cold-cured cementitious stone artefacts do seem to possess a very variable survival quality which is not entirely due to environmental effects and their initial moisture content can certainly be attributed to a major cause of their deterioration: the use of iron struts or armatures to internally strengthen the Austin & Seeley artefacts is also very susceptible to variable moisture content of the adhering composite as in some cases the struts have rusted and mechanically cracked the cementitious layers surrounding them, also staining these green and brown with iron oxides, iron hydroxides and other iron salts.

4.8 Was the Claim That Coade Stone Survived Better Than Natural Stone in the Environment Justified?

Eleanor Coade used this statement to great effect in marketing her artificial stone products and it was echoed by many of her architects and landscape gardeners, but in hindsight we should examine if it was justified? Certainly, at the time it was believed and the initial evidence for the correctness of this statement seems incontrovertible. For the manufactory production run over almost 70 years it was never seemingly challenged, so presumably there must have been some observable substance to the claim inherent at the time? Reference is made to Eleanor Coade's first success that established her statuary business very early on with Robert Adam, who was then in his ascendancy as an architect, when he commissioned several pieces in Coade stone for an ornamental archway fronting Syon House, the London home of the Duke of Northumberland. These were designed to replace the existing stone statuary, and it is alleged inferior artificial stone substitutes, that had been installed on this structure just a year or two hitherto and which had already deteriorated badly in comparison with the surrounding stone. The Coade stone pieces at Syon House are still there and in good condition even though it can be seen visually that the natural stone components in the structure surrounding them has now in comparison suffered rather badly. These pieces would have been manufactured during the first year's full operation of Eleanor Coade's factory, so it is clear that the correct recipe and formulation for their environmental survival had already been achieved by that time. Eleanor Coade was upset to learn from Robert Adam that several observers thought that the earlier defective artificial stone items that needed replacement at Syon House were her products also and she made strenuous public denials of this false information—it seems that George Davy and his cold-cured cementitious alternatives were responsible.. The website for English Heritage also refers to several grand houses that have now fallen into disrepair and have then been demolished, but noting that the paterae, keystones and friezes in Coade stone which formed part of them originally still survive in good to excellent condition when found amongst the rubble of the original structures. An example of this is provided by the estate of General Lord Robert Manners of Bloxholm House, Kesteven, Lincolnshire; although the house was demolished in 1963, the surviving Coade stone coat-of-arms, unicorn supporters and crest, inscribed Coade & Sealy, Lambeth, which was commissioned in 1772, had been removed and installed in the early nineteenth century over the porch of Bloxholm Church, where it now stands with its motto ("*Pour Y Parvenir*", meaning "*Strive to Attain*"). There is some damage observed, which can be attributed to mechanical stress during the re-siting process. Eleanor Coade made coats-of arms and these feature as items in her Coade & Sealy Catalogue of 1784, but only those of the Royal Family are cited there specifically, as these heraldic items are specifically very personal items to family arms-bearers. Philip Thomasson of Cudworth has spent a large proportion of his working life conserving and restoring Coade stone artefacts and he is of the considered opinion from working on more than two hundred different examples that they do indeed vary in quality from site to site despite there as yet being no analytical compositional data to confirm this. Whereas, it could be argued that local environmental issues could play a role in the survival of individual pieces, Philip Thomasson believes that the actual fabric of the artificial stone could perhaps be somewhat variable and plays an integral part in its eventual survivability. In some cases, physical damage to the artificial stone piece has clearly contributed to its overall demise by exposing inner chambers to atmospheric and terrestrial contamination, especially if metal struts (bronze or ferrous armatures) have been used to preserve the shape from the moulds as these are then subject to corrosion as has been found in cementitious examples, and vandalism has also played its part in exposing substratal domains within the statues. An interesting sequel of relevance relates to some garden statues from Sir John Soane's projects which display an inner reinforcement of Coade stone ceramic reinforcing bars rather than the use of metal struts: these have not shown signs of corrosion at all and this probably means that both methods (*viz.* metallic and ceramic) could have been adopted for strengthening and reinforcement of complex and larger Coade stone statuary (Kelly, *Mrs Coade's Stone*, 1990). Analytically, it should be possible to differentiate between the compositional variability of Coade stone statues and artefacts and perhaps identify sites of mineral damage instituted by environmental degradative issues which may then be correlated with the presence of specific raw materials, minerals or firing conditions, but this would require a major effort to sample partially degraded statuary for SEM/EDAXS or XRF microanalysis. Until scientific evidence can be brought to bear on this problem, the conclusion must remain, therefore, that there could be a distinct possibility that not all Coade stone artefacts may have the same production quality and that perhaps they did necessarily possess a consequently variable resistance threshold to environmental degradation issues.

4.9 Is All Coade Stone Correctly Identified?

This challenging statement covers several options:

- Are all Coade stone artefacts currently listed truly made of Coade stone, i.e. they have been made by the manufactory of Coade /Coade & Sealy and their successors at King's Arms Stairs, Narrow -Wall, Lambeth, between 1769 and 1837/1840?

- Has statuary made by the Coade successors Blashfield and Blanchard using the actual Coade & Sealy moulds purchased in 1843 and copies made from existing Coade statues which have been manufactured to the same recipe and formulation for the paste been accurately identified? It is clear that some examples of these later "Coade stone" artefacts have been marked appropriately but it is also possible that unmarked exemplars exist which could be confused with the earlier genuine Coade examples . In the case of genuine Coade stone artefacts, only some 20% or so were appropriately marked with an impressed stamp, some of which were also dated appropriately.
- In some well-known and documented cases, Blashfield and Blanchard, made secondary moulds of existing genuine Coade statuary and created reproduction copies in Coade stone, which sometimes bear their own factory marks. Can one distinguish between the genuine first-grade Coade statues and the later fired reproductions of Blashfield and Blanchard? It should be realised that the production of a secondary mould from an existing Coade stone statue implies that the eventual copy would be significantly smaller than the original because of thermal contraction of the paste during the firing process, which amounts to about 13%: Eleanor Coade allowed for this contraction for cases where precise sizes were specified by the client.
- Copies of Coade statues were also made in cementitious artificial stone which are contemporary with the Blashfield and Blanchard artificial stone reproductions, such as the pineapple from Ham House, Surrey, manufactured by Austin and Seeley. Analytically, it should be possible to differentiate between these cementitious and ceramic versions by their calcia, CaO, content, which for the cementitious statues will be found to be near 30% and for genuine ceramic fired Coade stone statues estimated to be less than about 1%.
- Alison Kelly (*Mrs Coade's Stone*, 1990) has complied a gazeteer in which she has identified over 700 sites of Coade stone artefacts in the United Kingdom and several overseas, many of which have multiple exemplars. Quite a few of these are supported by documentary evidence which establishes their forensic provenance but others are not and this must raise a warning flag concerning their true attribution. For example, are these more modern replacements or perhaps mis-assigned earlier reproductions?
- Conversely, there are instances it seems where the historical documentation asserting that an artefact is made of Coade stone is patently incorrect and, therefore, it is critical that every allegation or certification is checked for its correctness of attribution. In the analogous case of porcelain manufacture, the breakage or loss of items in contemporary dinner, tea and dessert services was sometimes occasioned with replacement by items from other factories, with these rather special and potentially unique items decorated *en suite* to match the originals and there are recorded examples of this occurrence. A parallel scenario for Coade stone would be the replacement of a damaged statue after closure of the Coade works by a similar one made from the same moulds by Blashfield or Doulton—technically this statue is no longer an original Coade stone artefact. Alison Kelly also noted several times in her classic text (Kelly, *Mrs Coade's Stone*, 1990) that local historians had documented what appeared to them to be Coade stone artefacts but which in reality upon closer examination were not.

4.10 Coade Stone Was the First Synthetic Fired Material to Be Successfully Used for Garden Statuary

This statement is only partially correct as some years before Eleanor Coade launched her artificial stone business and made literally hundreds of items of garden statuary, King Augustus II (The Strong), the Elector of Saxony, and prime motivator behind the Meissen porcelain manufactory in Dresden when he engaged Ehrenfried von Tschirnhaus and Johann Bottger to make his "*white gold*" in the form of Meissen hard paste porcelain which first appeared in 1711, had the idea of manufacturing large animals in porcelain. The Meissen factory was entitled at that time the Royal Polish and Electoral-Saxon Porcelain Manufactory (Koningslich-Polnische und Kurfurstlich-Sachsisische Porzellan Manufaktur) and was based at Augustus' castle in Albrechtsburg at Meissen, near Dresden. Augustus called his porcelain material "*white gold*"—this nomenclature directly reflects the original purpose of Augustus bringing together these two experimental chemists to work for him, which was to assist in the alchemical transmutation of base metals into gold (Keating, *The Hunt for White Gold*, 2011; Cassidy-Geiger, *Meissen and St Cloud, Dresden and Paris*, 1999). In 1728, King Augustus created his Japanese Palace to house his large porcelain collection of Chinese and early Meissen wares and he commissioned Johann Gottleib Kirchner and Johann Joachim Kaendler to model almost 600 animals and birds, of which 458 were actually achieved in the medium of porcelain by 1735. Many of these comprised the largest pieces of porcelain created at that time, some greater than life-size, and necessitated compositional changes in the Meissen porcelain body to withstand the firing conditions in the kilns (Wittwer, *A Royal Menagerie—Meissen Porcelain Animals*, 2006). Although not specifically designed as garden statuary and they were originally placed on the third floor of Augustus' Japanese Palace, later acquisitions were certainly used as ornaments in formal gardens . As referred to above, hard paste porcelain is similar in composition and in firing conditions to Coade stone and it is conceivable that Augustus' much admired porcelain menagerie statues could well have generated the idea of artificial stone statuary to Eleanor Coade and her predecessors some thirty years later in imitation of marble sculptures. In the same context, John Dwight established his Fulham Pottery in London in the 1670s with assistance from the Royal Society, who were interested in the synthesis of high temperature refractory crucible materials for the extraction of metals from their ores. Dwight certainly made porcelains and quality stonewares using high temperature kilns at Fulham and is recognised as the first person to manufacture ceramic figures: his representations of Prince Rupert of the Rhineland-Palatinate (about 1700 or perhaps earlier) and his daughter Lydia Dwight (who died, aged 6 in 1674) are now preserved in the British Museum and the Victoria and Albert Museum, London, respectively, and are remarkable for their lifelike appearance and design detail quality.

4.11 Eleanor Coade Patented Her Coade Stone

It may seem surprising now but Eleanor Coade never patented her Coade stone: this may seem obvious in the light of our knowledge about the method of her acquisition of the business and that she probably effectively used Daniel Pincot's recipe and formulation to achieve this (Pincot 1770), which he had not previously patented, although he claimed correctly that it was sensibly different from that of Richard Holt's patent of 1722. In 1730, Holt made a legal deposition in which he revealed his recipe and it was very different from that recipe used by Eleanor Coade some years later (Holt, 1730 and 2010). It is quite possible at that stage in 1769 that Pincot's formulation was already well known and it could also be that variants were being used elsewhere or that patent breaking for artificial stone formulations was then relatively easy because so much knowledge was already available to practitioners in the public arena. This scenario would also lend credibility to the hypothesis that Eleanor Coade did not amend Pincot's recipe and formulation significantly when she took over his company in 1769 as that would obviously need some further protection from erstwhile competitors to a shrewd businesswoman. Yet, several historians have assumed incorrectly that such a "novel material" would have been patent protected and the first sentence of the introductory foreword to Alison Kelly's authoritative survey of Coade stone published in 1990 (Kelly, *Mrs Coade's Stone*, 1990) actually contains this premiss:

> The artificial stone patented by Mrs Eleanor Coade may not have revolutionised the architecture of the late Georgian period in any structural sense but its aesthetic contribution was much more considerable than is generally realised; paradoxically, indeed, its popularity amongst architects and clients was in direct proportion to its subsequent anonymity, since its success in imitating the real thing was the reason both for its widespread use and for the fact that modern writers have tended to ignore its existence ,,,,,, Coade stone made it possible for architects of the greatness of Soane and Wyatt to exploit, develop and deploy the infinite variety of the classical vocabulary of ornament in a way that reliance on natural stone would have rendered impractical (Roger White, The Georgian Group, London, 1990).

Incorrect though the first remark concerning Eleanor Coade's patent is, the following statement about the realistic quality of the artificial Coade stone to observers and the enhancement it provided to architects and landscape gardeners for their statuary and building embellishments is very true. Alison Kelly echoes this thought later in her book:

> Her clay imitations are so realistic that people think they are the real thing. Camouflage is perfect! How many looking up at the fan vaulting of St George's Chapel, Windsor, actually realise it is artificial stone and not the real thing – a roof of pottery?

4.11.1 Another Reference to Eleanor Coade's Patent?

Following on from the misconception generated in the foreword to Alison Kelly's book discussed above, a further instance of this is made in the transcript below from a paper delivered by the Rev. V. Hope on the Templer Family of Somerset, Devon and Dorset delivered on the 14th January 1974:

> About 1769 Mrs Coade settled in Lambeth, her premises being at the Pedlar's Arms opposite Whitehall Stairs. She seems to have acquired the patents and buildings of Richard Holt. By 1811 the manufactory had become the property of Messrs Coade and Seeley (the latter being Mr Coade's nephew). On Seeley's death a William Croggan (clerk or manager) became the proprietor. About 1827 Croggan removed the business to the New Road near Tottenham Court.

Within the Rev. Hope's presentation, reference was made to Thomas Pennant's work (*Some Account of London*, 1790) which gave his personal history of the growth of the city from Roman times and mentions Mrs. Coade's artificial stone manufactory:

> In a street called Narrow Wall is Mrs Coade's manufactory of artificial stone . . . the statue, the vase, the urn, everything that could be produced out of natural stone or marble is here to be obtained as an easy rate.

Pennant was born in 1726 and died in 1798 and his book appeared in several editions up to 1813. There are several incorrect statements of fact in the Rev. Hope's paper: Eleanor Coade is not mentioned specifically and there seems to be some confusion between Mrs. Coade and Mr. Coade, she certainly did not acquire the patents of Richard Holt or his manufactory at the Pedlar's Arms, Seeley (note the mis-spelling!) never owned the manufactory and William Croggan (note the mis-spelling!) certainly did not become the proprietor after John Sealy had died in 1811.

References

J.A. Anderson, *Derby Porcelain and the Early English Fine Ceramic Industry*. Ph.D. thesis (University of Leicester, UK, 2000)

M. Cassidy-Geiger, Meissen and St Cloud, Dresden and Paris, in *Discovering the Secrets of Soft Paste Porcelain at the Saint Cloud Manufactory, ca. 1690–1766*, ed. by B. Rondo, (Bard Graduate Center for Studies in the Decorative Arts/Yale University Press, New York/New Haven Connecticut/London, 1999), pp. 97–111

R. Denyer, M.C. Denyer, H.G.M. Edwards, *Armorial Porcelains: The Genesis* (Springer-Nature, Dordrecht, 2024)

H.G.M. Edwards, *Porcelain to Silica Bricks: The Extreme Ceramics of William Weston Young, 1776–1847* (Springer -Nature, Dordrecht, 2019)

H.G.M. Edwards, *Porcelain Analysis and Its Role in the Forensic Attribution of Ceramic Specimens* (Springer -Nature, Dordrecht, 2021)

K.A. Esdaile, Coade stone. Archit. Build. News **CLXI**, 94–96. and 112–114 (1940)

J. Fairbairn, *Crests of the Families of Great Britain and Ireland and Their Kindred in Other Lands, Volumes I and II* (T.C. and E.C. Jack/New Orchard Publishing, Edinburgh/London, 1905 and 1986)

C. Fowler, Some remarks on terracotta and artificial stone as connected with architecture. Civil Eng. Archit. J. **13**, 215–216 (1850)

J. Foyle, How a Sculptor Cracked the Lost Mixture of Eleanor Coade's Stone. *Financial Times*, London, July 15th, 2015

I. Freestone, Forgotten but not lost: the secret of Coade stone. Proc. Geol. Assoc. **102**, 135–138 (1991)

R. Holt, *A Short Treatise on Artificial Stone as 'tis Now Made and Converted into All Manner of Curious Embellishments and Proper Ornaments, of Architecture, London, 1730*; reprinted by Gale ECCO (Eighteenth Century Collection Online), Detroit, Michigan, USA, 2010

M. Keating, *The Hunt for White Gold* (Hodder & Stoughton, London, 2011)

A. Kelly, *Mrs Coade's Stone* (Self-Publishing Association, Worcestershire, 1990)

T. Pennant, *Some Account of London* (Robert Faulder, London, 1790)

D. Pincot, *An Essay on the Origin, Nature, Uses and Properties of Artificial Stone* (Richard Hett, London, 1770)

H. Roberts, W.H. Godfrey, *Survey of London: Lambeth: South Bank and Vauxhall* (London, 1951)

J.E. Ruch, Regency Coade: A Study of the Coade Record Books, 1813–1821. Archit. Hist. **11**, 34–36 (1968)

S. Wittwer, *A Royal Menagerie- Meissen Porcelain Animals* (Getty Trust Publications/The J. Paul Getty Museum, Los Angeles/The Rijksmuseum/Amsterdam, 2006)

Chapter 5
The Mineralogy of Fired Ceramics

Abstract The mineralogy of the raw materials and of the fired ceramics used in the production of artificial stone and the species that are formed form their fusion at high temperatures in the kiln are discussed along with details given of the chemical changes that occur during the firing process at several temperature ranges. The sourcing of the raw materials and of their chemical composition are presented and considered for primary and secondary clays sand, flint, alkaline flux, glass frit, grog and feldspar are outlined. The analytical interrogation of the minerals formed in ceramics at high temperatures in the kiln are described.

Keywords Clays · Sand · Flint · Grog · Glass frit · Alkaline flux · Feldspar · Unaker · Kaolin · Mineral identification using elemental and molecular spectroscopic analysis

In common with their porcelain analogues, the raw materials used in the preparation of the artificial stone pastes that are then fired in the kiln at high temperature do not possess the same mineralogy afterwards and it is appropriate here to discuss the chemical changes that can occur to these minerals during the firing process. Although the determination of the elemental oxide composition of the fired ceramic is essentially unaffected by the thermal treatment imposed upon the raw materials, i.e. the silica, alumina and other elemental oxides are not created or destroyed in the firing process, the molecular formulations of the original materials at high temperatures takes on new phases and these then afford an opportunity for the application for a complementary molecular spectroscopic analysis to determine the presence or otherwise of key mineral signatures that can be used to estimate the temperatures achieved in the furnace or kiln during the firing process. Chemical changes in minerals and compounds occur thermally with the elimination of volatile species above critical temperature ranges, such as carbon dioxide from carbonates, the loss of bound water of hydration, nitrogen oxides from nitrates and sulfur oxides from sulfates. Here now follows a consideration of the key chemical changes and reactions that can be expected to occur in the firing of artificial Coade stone paste after it has been pressed into the moulds and shaped into the desired form.

H. G. M. Edwards, C. J. Brooke, *Coade Stone*, Cultural Heritage Science,
https://doi.org/10.1007/978-3-031-71432-0_5

5.1 The Raw Materials in Artificial Stone Synthesis: Their Sourcing, Mineralogy and Chemical Composition

The raw materials that have been used to manufacture artificial stone and stoneware ceramics historically seem to be straightforward to appreciate but the area can be compromised by the incorrect usage of descriptors and by an often rather loosely applied terminology which compounds the problem of precise identification of the material used for later researchers (Edwards, *Porcelain Analysis and Its Role in the Forensic Attribution of Ceramic Specimens*, 2021). For example, Lewis Dillwyn FRS, proprietor of the Swansea China Works, as a scientist recognised the ambiguity of the commonly used descriptor for *alabaster* that was applied indiscriminately to both calcium sulfate dihydrate, $CaSO_4.2H_2O$, and to calcite, $CaCO_3$, by ceramics manufacturers with often woeful results ensuing for their production process. Chemically the two calcium salt minerals are different but the idea that alabaster gypsum is a form of calcium carbonate has been perpetuated into modern times: Hettie Judah in her recent book (*Lapidarium: The Secret Lives of Stones*, pp. 187–191, 2022) alludes to their interchangeability and it appears that the distinction of usage description between alabaster and calcite depends upon whether one uses an archaeological approach or a geological one. The most difficult concept for early ceramics manufacturers to understand was the nature of the thermal reactions that were undergone by their raw materials in the kiln upon reaching the elevated operating temperatures of the kiln in their processes: the manufacturers certainly had the correct idea that they should use starting materials of the highest purity available to them at that time but they were unable to control other key factors, some of which would have been recognised by them empirically through the practical knowledge acquired from their predecessors. A brief coverage of the commonly used raw materials in artificial stone synthesis by Eleanor Coade and her successors will now follow and some of the pitfalls that can exist in the interpretation of the recipes that were used by them will be highlighted.

5.1.1 Clay

The term "clay" is geologically very indefinite and is an excellent example of a general terminology which covers many chemical possibilities from the basic primary clays (formed by the *in situ* geological degradation of granites and feldspars) and secondary clays (formed as sediments after their hydrological transportation and eventual deposition some distance from their source) which themselves can also differ in their chemical composition depending upon their geographical sources. Clays do possess some more specific descriptors which can be found in the general ceramics literature, such as *ball clay*, *pipe clay*, *china clay* and *kaolin* but these can actually hide several distinct possibilities. All clays are hydrous aluminium phyllosilicates chemically and some have precise chemical mineral formulations, whereas others are of an indefinite chemical composition that can arise because of their ability as minerals to intercalate other metallic ions from their surrounding geological environments, such as Mg^{2+}, Ca^{2+}, Fe^{2+}, Fe^{3+}, Na^+ and K^+ and many others too, such as Bi^{3+}, Ba^{2+}, Ti^{4+}, Ni^{2+}, Zr^{4+}, Rb^+, $Sr^{2+}$and Mn^{4+}, which can be detected using modern analytical equipment. Primary clays are generally formed geologically from the degradation of volcanic ash and basalts upon exposure to marine environments and primarily are formed through the hydrothermal weathering of basalts emerging from undersea volcanic vents. In contrast, secondary clays are formed geologically from the further terrestrial hydrothermal weathering and hydrological transportation of the decomposition products of granite and feldspars.

Professor Victor Owen in an excellent article (Owen 2002) has summarised this structural situation in particular for clays, wherein he has indicated that clays can actually be divided into five distinct groups of minerals, namely, *kandites* (e.g. kaolinite, bentonite), *illites* (e.g. illite, celadonite), *smectites* (e.g. montmorillonite, saponite), *palygorskites* (e.g. sepiolite) and *vermiculite.* In other geologically themed articles, vermiculite is itself defined as a smectite and not as a separate entity. Another way of defining clays is based upon the tertiary structural assembly of their two-dimensional sheets of silicate (Si-O) and aluminate (Al-O) components. The prime structural feature of a clay involves the three-dimensional linking of sheets of tetrahedral SiO_4 units to sheets of octahedral AlO_6 units through strong O...H-O hydrogen bonded entities or additionally involving the more weakly-bonding H_2O molecules that are intercalated between the individual sheets. Generally, at room temperature the intercalating weakly bonded H_2O molecules permit an easy slippage to occur between the sheets of silica and alumina, which gives the clay its plasticity when wet: after being fired, however, these weakly bound water molecules have been removed and the stronger interlayer hydrogen bonds then take over the control of the ceramic structure, which gives the fired clay artefact its rigidity and strength. There are three different physico-chemical descriptions of these intersheet layer structures in clays, namely, silica: alumina in 1:1 and 2:1 symmetric arrangements and also in a random arrangement. Primary clays such as *kaolinite* have a symmetric 1:1 silica: alumina layer structure which is often paraphrased as T-O (Tetrahedral-Octahedral, see above), whereas secondary clays such as *illite*, *montmorillonite* and *talc* have a 2:1 symmetric arrangement written as T-O-T, where T refers to the tetrahedral structure within the layer and O is the octahedral intralayer structure. The third type of clay structure is the random asymmetric arrangement of tetrahedral and octahedral layer arrangements exemplified by groups of clays known as the *allophanes*, which have a general formulation of $Al_2O_3.2SiO_2.H_2O$ or less specifically as $Al_2O_3(SiO_2)_{1.3\text{-}2}2.5\text{-}3H_2O$. A fourth type of clay structure actually exists in terms of this structural arrangement of adjacent layers and is exemplified by *chlorite*, which is represented formulaically as $(MgFe)_3(SiAl)_4O_{10}(OH)_2(MgFe)_3(OH)_6$, having a 2:2 layer arrangement wherein an $Mg^{2+}$and Fe^{2+}octahedral layer is interspersed between two silica tetrahedral layers and an alumina octahedral layer. *Chlorite* is normally found to occur along with *brucite*, which supplies the magnesium hydroxide in an octahedral coordinated metal form suitable for its direct structural incorporation into the clay structure. A list of these clay types and their classification is provided in Table 5.1.

Table 5.1 Elemental oxide compositions/% of raw materials used in fired artificial Coade stone

Raw material[a]	SiO_2	Al_2O_3	CaO	MgO	Na_2O	K_2O	PbO	Fe_2O_3
Sand (10%)	99							1
Flint (10%)	100							
Glass frit (10%)	60	20	4	2	10	5		
Ball clay (50%)[b]	55	40	1	1	1	3	1	
Kaolinite (50%)								
Mica (20%)								
Quartz (30%)								
Grog (20%)	40	40	2	4	4	4	4	
Ball clay (65%)								
Quartz (15%)								
Feldspar (15%)								
Coade Stone predicted[c]	61	30	1.3	0.7	1.7	2.1	0	1.3
Coade Stone observed[d]	57	30	1	1	1	3	0	2

[a]The figure in brackets is the % composition of this component in the Coade stone recipe published by Charles Fowler in 1850

[b]The analytical data of Ian Freestone (1991) from a single Coade stone specimen is stated to be remarkably similar to the composition of the original ball clay used in its raw materials

[c]The predicted Coade stone composition analytically based upon the recipe and mineral compositions

[d]The observed analytical composition of Coade stone (Brooke et al. 2022)

Many clays have metal ions which have been incorporated into their host silica and alumina layer matrices from their surrounding environments: for example, vermiculite has taken Mg^{2+} and Fe^{2+} in substitution for Al^{3+}, montmorillonite has Mg^{2+} replacing Al^{3+}, biedellite has Al^{3+} replacing Si^{4+}, nontronite has Fe^{3+} replacing Al^{3+}, and saponite has Al^{3+} and K^{+} replacing Si^{4+}. Often these ionic substitutions result in a charge imbalance in the clay and this affords the additional possibility of the take-up and absorption of further ions into the clay from the environment: this property gave rise to the common practice adopted by ceramics manufacturers of washing their clays thoroughly before use as their raw materials to leach out some of these unwanted adsorbed ions that it was recognised could interfere with the outcome of the firing process, particularly for the removal of excess iron oxide or salt deposits which might colour the resulting ceramic substrate upon firing. In particular, *ball clays* can comprise several different types of clay and their uptake of metal ions can be reflected in a wide range of colours: the "blue clays" used in English ceramics manufacture were ball clays from Dorset, in particular the Isle of Purbeck and the Wareham Basin, which were prized for their incorporation in admixture along with kaolinite to confer a desirable plasticity on the raw material compounded porcelain paste for its subsequent moulding into artefacts at the manufactory.

Often, a mixture of primary and secondary (ball) clays was adopted by manufacturers to facilitate a greater plasticity in the throwing of the pottery or moulding of the paste. Seemingly Eleanor Coade did not do this and used exclusively Dorset ball clay. Again, it would have been almost impossible for ceramics manufacturers to predict the composition of their clays before the advent of chemical analysis, because after all these are natural geological mixtures of raw materials, but the proprietors circumvented this obvious problem by using the same mines wherever possible for their source of the clays. When the particular mining operation at a specific site ceased to function or when the deposits became worked out it generally proved to be a disaster for the ceramics industry concerned and a notable case of this occurred at the Limehouse Pottery when its preferred source of raw material from Kynance Cove in Cornwall became worked out in mid-1747 and by early 1748 the manufactory had been forced to close down (Ramsay et al., *The Limehouse Porcelain Factory: Its Output, Antecedents and the Influence of the Royal Society of London on the Evolution of English Porcelain Based on Composition and Technology*, 2013). An interesting tale in this context emerges around the person of Andrew Duché, who noticed in the early years of the eighteenth century that the Cherokee Indians in the vicinity of Charleston, South Carolina, where he was based, had adopted the use of a very fine quality local white clay (*unaker*) for making their pottery. He investigated and was impressed by what he saw of their earthenware manufacturing process, eventually taking specimens of the Cherokee clay and bringing them to London, where it is alleged that he managed to interest the founding proprietors of the Bow porcelain manufactory, Edward Heylyn and Thomas Frye, in its use for their manufacture of their earliest porcelain (Ramsay and Ramsay 2008). Heylyn and Frye immediately took out a patent for this Cherokee clay, called *unaker*, which was not duly credited as feasible by several modern observers until Ross and Gael Ramsay went to the Carolinas and sought out themselves the source of this ephemeral clay, which indeed does exist and analyses as a very high quality kaolin indeed! So, this tale illustrates that even before the days of analytical chemical science, the appearance and sourcing of a fine white "potter's clay" could be appreciated by the founders of the ceramics manufactories.

Ball clay is not a precise description of a single mineral and in reality it comprises a mixture of kaolinite (white china clay), mica and quartz, as indicated in Table 5.2; the importance of using the identical raw material sourced from the same mine is now seen to be an essential requirement for production purposes as this was the only way manufacturers could have of standardising their raw material without the benefit of chemical analysis. Chemical analytical procedures for the qualitative and quantitative estimation of the elemental and molecular composition, and the consequent purity, of a raw material were not available until the early nineteenth century following the researches of Lavoisier, Priestley, Reaumur and others (Edwards, *Porcelain Analysis and Its Role in the Forensic Attribution of Ceramic Specimens*, 2021). The main advantage of the use of a ball clay over purer forms of kaolinite is that it afforded a better plasticity for moulding of the ceramic paste before firing. It was also especially low in iron oxide impurities and this resulted in the retention of the most desirable white or cream body texture after firing.

Table 5.2 Chemical formulation of minerals in the raw materials of Coade stone

Raw material	Mineral	Source	Chemical formula	Contribution to elemental oxides
Silica[a]	Quartz	Reigate sand	SiO_2	SiO_2, Fe_2O_3
Dorset Clay[b]	Kaolinite	Purbeck/ Wareham	$Al_2Si_2O_5(OH)_2$	SiO_2, Al_2O_3
	Mica		$(K, Al, Na, Mg, Fe)_2$ (Si, Al, Ti$)_3$ $O_{10}(OH)_2$	SiO_2, Al_2O_3, K_2O, Na_2O, MgO, Fe_2O_3, TiO_2
	Quartz		SiO_2	SiO_2
Flint	Quartz	East Anglia	SiO_2	SiO_2
Glass[c]	Aluminosilicate	London	$Na_2O.Al_2O_3SiO_2.CaO.$ K2O	Na_2O, Al_2O_3, SiO_2, CaO, K_2O, PbO
Grog[d]	Ball clay	Purbeck/ Wareham	$Al_2Si_2O_5(OH)_2$	SiO_2, Al_2O_3
	Quartz	Reigate sand	SiO_2	SiO_2, Fe_2O_3
	Feldspar[e]	Devon/ Cornwall	$KAlSi_3O_8$	K_2O, Al_2O_3, SiO_2
			$NaAlSi_3O_8$	Na_2O, Al_2O_3, SiO_2
			$CaAl_2Si_2O_8$	CaO, Al_2O_3, SiO_2

[a]The source of the sand provides the variability in the iron(III) oxide content: the finest and whitest river sand generally contains significantly less than 1% of Fe_2O_3, and this was targetted by Eleanor Coade as it resulted in the creamy whiteness of the fired ceramic body

[b]Dorset ball clay is a mixture of kaolinite, mica and quartz, which has been deposited in secondary sediments from the degradation of granite and feldspars. It was highly acclaimed on two counts for ceramics usage: firstly, it afforded a better plasticity when compared with kaolinite clays alone and secondly, it resulted in a white ceramic body after firing. The small percentage signatures of magnesia and titania seen in Coade stone analytically are attributed to impurities arising from this source

[c]Glass is of an indefinite composition and its elemental content is usually outside the control of the ceramics proprietor: it was usual to purchase "end-of-day" glass from local glassworks and this was compounded of different glass mixtures that had been used during the day. Normally there are two types of glass: crown glass, which contains soda, potash, alumina and sometimes calcia, effectively a sodium aluminosilicate, but the alternative was the more expensive flint glass which contained added lead oxide ranging between 10 and 60% which was used to make the best quality lead crystal glass for engraving and etching. The presence of a lead signature in a fired unglazed ceramic is indicative of the use of lead glass frit as a raw material

[d]Grog, or *chamotte,* is also of a rather indeterminate composition dependent upon it sourcing. In its widest descriptor it is fragmentary waste material from previous ceramic firings, which may contain glazed shards, but Eleanor Coade did specify that she preferred using stoneware grog, which was a high-temperature fired ceramic that closely resembled her own artificial stone formulation. Several local manufactories would be making white-bodied stonewares and this raw material would have been purchased cheaply in ton quantities for fine grinding in her Lambeth manufactory

[e]Feldspar is a generic term for a group of geological minerals based upon metal silicates such as biotite, $K(Mg,Fe)_3(Al,Fe)$ $Si_3O_{10}(OH)_2$, and muscovite $KAl_2Si_3AlO_{10}(OH)_2$: here the three forms of commonly encountered feldspars in ceramics manufacturing, orthoclase $KAlSi_3O_8$, albite $NaAlSi_3O_8$ and anorthite $CaAl_2Si_2O_8$, are given

5.1.2 Sand

Generally, ceramics manufacturers would seek to use the finest white quartz river sand which was low in its iron oxide content, Fe_2O_3, as this would otherwise confer a yellow to red cast upon the fired ceramic body, as it does for terracotta. Others would add a small quantity of *smalt*, a synthetic cobalt oxide glass, which is blue in colour and often represented chemically as $Co_2O_3.nSiO_2$, which reflects its variable chemical composition—normally smalt would contain about 10% cobalt oxide (Jonyunaite et al. 2010); the blue smalt would effectively remove the appearance of a yellow cast colour from the fired ceramics attributed to the presence of iron (III) oxide. Note that *smalt* is a term which is often used interchangeably with another synthetic blue glass, *Bristol blue*, which is a cobalt aluminosilicate, represented as $Co_2AlO_4.nSiO_2$ (Edwards, *Porcelain Analysis and Its Role in the Forensic Attribution of Ceramic Specimens*, 2021). *Smalt* could comprise a compositional range which varied between 10% and 15% of cobalt oxide. River sand was the preferential choice for a raw material as it is already finely divided and could be used directly in the preparation of an initial frit: the transportation of fine white river sand to a manufactory site over some considerable distance was often undertaken. For example, William Duesbury of the Derby China Works imported selectively river sand from East Anglia and flints from either the Isle of Wight or East Anglia (Anderson 1991, 2000). The manufacture of earthenware did not impose such a stringent requirement upon the raw materials and the characteristic terracotta red colour of earthenwares was due to the presence of significant amounts of iron oxide in the local sand sources used. Stonewares and artificial stone on the other hand fall between porcelain and terracottas in this respect and generally higher quality river sands were used in their manufacture. Eleanor Coade certainly subscribed to its use as a raw material for her Coade stone, specifying fine Reigate sand in her recipe.

5.1.3 Flints

The use of flints, chert or chalcedony, a highly pure form of silica, which occurs as nodules in sedimentary deposits, was advocated in many ceramics manufactories; again the Lynn gravel beds of East Anglia and sources in the Isle of Wight were particularly favoured by ceramicists as a source of their flint raw material.

5.1.4 Alkaline Flux

The addition of an alkaline component in the ceramic paste mixture to promote fusibility and achieve vitrification at high temperature was well known to porcelain proprietors. The inclusion of calcite, dolomite or dolomitic limestone as a raw material component to provide a local contribution of alkaline calcia or magnesia upon heating in the kiln was used by some manufacturers but was deplored by others, such as Lewis Weston Dillwyn at Swansea, on account of the generation of gas bubbles of carbon dioxide that would be released in the thermal chemical decomposition of these minerals in the kiln which would detract from the translucency of the resulting ceramic—and this decrease in visual light transmission was a disadvantage for the fired products in porcelain synthesis where translucency was desirable. This was not so vital for stonewares and Eleanor Coade did not specifically add lime to her paste mixtures for artificial stone, although it has been noted that on occasion a slip of limestone and lime was possibly added to the outer surface of her Coade wares. Most ceramics manufacturers also added an alkaline flux in the form of soda ash and/or potash, which were obtained from the thermal decomposition of terrestrial or marine plant and vegetable matter or seaweed. The ash formed from the burning of terrestrial plants is potassium rich whereas that obtained from the burning of kelp or seaweed is sodium rich: the soda ash and potash component derivatives are, therefore, necessarily of a rather indefinite composition and were very dependent upon the source of the calcined ash, both containing soda and potash in varying relative amounts. Although potash chemically is written formulaically as potassium carbonate, K_2CO_3, and soda ash as the sodium analogue sodium carbonate, Na_2CO_3, these are idealised chemical formulations of a very intractable raw material. Analytically, there was no means in the eighteenth or early nineteenth centuries for manufacturers to effectively control the chemical composition of this raw material, in common with many others that they had to use in their ceramic syntheses. We should also be aware that several alternative names were used for these raw materials at that time historically; for example, for potassium carbonate alternative names of carbonate of potash, salts of tartar, pearl ash and muriate of potash were used, the latter also contained a significant proportion of potassium chloride KCl. For sodium carbonate, also known as washing soda and simply soda crystals, there was a natural mineral alternative in trona or natron, which was obtained from saline lake deposits as a crystalline mineral formulated as $Na_2CO_3.NaHCO_3.2H_2O$ for trona, also known as sodium sesquicarbonate; this was mined as natron in ancient times by the Egyptians and Romans at Wadi El Natrun in the Western Desert in Egypt and used prolifically in their manufacture of glass and also as a desiccant for cadavers in their mummification and body preservation rituals. Natron is not as precisely formulated chemically as trona and contains about 83% $Na_2CO_3.10H_2O$ (washing soda) and 17% $NaHCO_3$ (baking soda), with small amounts of sodium chloride and sodium sulfate also being present. Another mineral that was occasionally used in place of soda ash was gaylussite, formulated as $CaNa_2(CO_3)_2.5H_2O$, which is effectively an equimolar mixture of calcium and sodium carbonates. Finally, soda ash actually contains sodium carbonate in several degrees of hydration, namely, $Na_2CO_3.H_2O$, $Na_2CO_3.7H_2O$ and $Na_2CO_3.10H_2O$, which are all converted to the anhydrous sodium carbonate, Na_2CO_3, upon heating to between 150 and 200 °C. The preferred alternative to the use of calcite as a flux was the addition of lime which had been prepared commercially using pulverised limestone as calcite $CaCO_3$, aragonite $CaCO_3$ or dolomite (dolostone) $CaMg(CO_3)_2$, which had been heated to around 900 °C, whereby the carbonates decomposed thermally to calcia or magnesia, CaO or MgO, or a mixture of the two. Lime manufacture was an important production of an industrialised society as it had several uses as an agricultural additive to acid soils, as a component of mortar and cements and in combination with chlorine as a bleach. The major problem was one of transportation as lime slowly absorbs water from the atmosphere exothermally to form slaked lime which then absorbs atmospheric carbon dioxide to reconvert it back into the carbonate. Depending upon the time that has elapsed between its manufacture and use in the ceramics industry, and especially considering its storage and transportation in wet weather, the amount of the active alkaline principle, which was either CaO or MgO, in the raw material could be very variable indeed and the amount of the carbonates formed in the stored lime before usage could build up to a significant level that would be detrimental to the integrity of the prototype ceramic in the kiln charge. The reaction of quicklime with either water or carbon dioxide in moist air has been discussed elsewhere here in terms of hydraulic cements.

It seems that Eleanor Coade relied upon the addition of glass frit, or cullet, to provide the alkaline fusion capability for her body paste as this raw material already contains derivatives of soda and potash from its manufacture and possibly a little calcia as well. Hence Coade stone presents a low calcia content upon analysis (~1%) in comparison with other contemporary ceramics.

5.1.5 Glass Frit

Another common additive for early stonewares and porcelains was glass frit, or cullet, which was generally purchased as the "end-of-day glass" from glass manufactories operating nearby: although ostensibly providing a source of silica and also of alkali metal oxides, its purity was inaccessible to manufacturers especially when the glass mixture could contain variable amounts of crown glass (soda and potash glass) and possibly the highly refractive lead oxide rich flint glass (Heinemann and Maggetti 2014). Analysts now find evidence of this variability in their interpretation of the lead content of early porcelains and glazes and its potential attribution arising from the deliberate addition of lead oxide (litharge PbO, or red lead, *minium*, Pb_3O_4) to the porcelain paste or to the inadvertent addition of flint glass as ground cullet, or both! Tite and Bimson (1991) recognised this particular difficulty when attempting to correlate the elemental determination of lead in a porcelain body or glaze with the amount of glass frit raw material used and Dr. Michael Tite and Dr. Mavis Bimson have estimated a typical median figure of 30% lead oxide would be present in the component for such a glass. Eleanor Coade was no exception in the use of glass frit for her artificial stone and this appears to be the main source of alkali metal oxides such as soda and potash for her flux as she did not use lime as a separate raw material additive. It is also clear that she preferred soda-lime glass, or crown glass, over the lead-oxide rich flint glass, although Richard Holt specified the addition of a significant amount of lead oxide in his recipe for artificial stone.

5.1.6 Grog

The other raw material of note that provided an additive for most stonewares, porcelains and artificial stone was grog: again, this is an example of an additive of indeterminate composition in the eighteenth and nineteenth centuries as it comprised porcelain shards and fragments of broken china or stonewares from other factories, usually purchased in ton quantities and transported by wagons to the manufactory site (Heinemann and Maggetti 2014). The reason for its specific use is somewhat of a mystery outside of it providing a cheap additive as the waste product of other manufactories, but in a powdered, finely ground or milled form it did provide a source of raw material components that had already been once-fired or twice-fired ceramically, then becoming wasters and being discarded or sold elsewhere. The major problem is that the manufacturers who were using grog from other ceramic sites literally had no idea of its chemical composition relative to their own wares and modern analysts need to be aware of its incorporation into the specimens they are investigating. For example, the incorporation through the grog component assimilation of key elements such as Mg, P, or Pb into ceramics where their normal manufactory usage did not comply with the published recipe can cause some ambiguities when the presence in the analytical data of such elements even in small concentrations is being discussed and compared with recipes from known examples of the genre for the same manufactory which did not perhaps incorporate precisely the same grog into their formulations. It is perhaps significant that some manufacturers did not use grog as a component in their wares for this reason: William Billingsley is not recorded as ever having used grog at Nantgarw as he was particularly concerned about the use of only the highest quality materials from designated sources and shards or fragments of interloper porcelains from grog, often providing the only archaeological evidence of their incorporation and usage in a manufactory, have not been found in archaeological excavations at the Nantgarw China Works site. Eleanor Coade, in contrast, used grog in her recipe (but does not indicate her source) and highlighted its incorporation into her artificial stone, along with the glass cullet referred to above, in her naming of her product as *lithodipyra* (Greek: twice-fired stone), which specifically acknowledges their previous high-temperature thermal history and their incorporation into her single-fired artificial stone paste! Some historians and chroniclers of artificial stone have assumed erroneously from the use of the word *lithodipyra* that Coade stone was itself twice-fired, resembling thereby the production of soft-paste porcelains, but in fact Coade stone was fired in a single process and is therefore similar to the production of hard paste porcelains (including the Chinese or "true" porcelain) and bone china. The historians were further convinced of the analogy between Coade stone and soft paste porcelain production in that the maximum operating kiln temperatures in each case were both in the region of 1150–1200 °C and that this was maintained for similar thermal exposure times of four days at this maximum temperature.

5.1.7 Feldspar/Steatite

The inclusion of steatite and feldspar in ceramic paste formulations would provide a highly siliceous basis for the body composition: steatite (soaprock, soapstone, talc) is a magnesium silicate, $MgSiO_3$, and generally the presence of magnesium oxide found in an analytical determination is sufficient to suggest that steatite has been incorporated into the ceramic paste mixture to give a *magnesian* ceramic which is more robust and texturally very different from the highly phosphatic soft paste porcelains that it replaced. Feldspar on the other hand resembles the *petuntse* used in the Chinese hard paste porcelain formulations: although *petuntse* is not found in the West, it is a feldspathic rock which is close formulaically to the *china stone* of Cornwall and other feldspars in chemical composition. There is no doubt that the discovery of kaolin clay and *china stone* feldspar in close proximity to each other at St Austell in Cornwall, which has been attributed to William Cookworthy, founder of the Plymouth hard paste porcelain manufactory in 1768, gave a major boost to English porcelain synthesis in the mid-eighteenth century (Church, *English Porcelain: A Handbook to the China Made in England During the Eighteenth Century as Illustrated by Specimens Found Chiefly in the National Collections*, 1894; Jewitt, *The Ceramic Art of Great Britain from the Prehistoric Times Down to the Present Day*, Vols. I and II, 1878). Ramsay and Ramsay (2008) give an excellent account of the differentiation between *petuntse* and the alternative forms of china stone used in porcelain synthesis and their importance in the first syntheses of English hard paste porcelains in simulation of the Chinese imports in the mid-eighteenth century and contemporary with Coade stone synthesis. Although Eleanor Coade did not specifically include a feldspar component in her artificial stone recipe, it could occur in her paste through the inclusion of grog, which was the product of stoneware factories and stonewares and certainly could have included feldspar as a raw material component. Feldspar is generically a crystalline component of granites and has a variable elemental oxide composition of metal silicates which is dependent upon its mineral content and sourcing. For example, minerals based upon metal silicates such as biotite, $K(Mg,Fe)_3(Al,Fe)\ Si_3O_{10}(OH)_2$, and muscovite $KAl_2Si_3AlO_{10}(OH)_2$ are classified as feldspars and three forms of commonly encountered feldspars in ceramics manufacturing are orthoclase $KAlSi_3O_8$, albite $NaAlSi_3O_8$ and anorthite $CaAl_2Si_2O_8$. Analytically, these would contribute to elemental oxide determinations through the detection of higher concentrations of soda, potash and calcia in Coade stone.

5.1.8 Minor Additives

Other minor additives were often invoked by ceramics proprietors to increase the plasticity of their ceramic paste, to reinforce the alkalinity of the mixture or to ensure that excess iron (III) oxide in the sand used did not materially affect the colour of the resultant fired porcelain or stonewares: these additives include borax, $Na_2B_4O_7.10H_2O$, white arsenic, As_2O_3 and smalt, a potassium cobalt silicate glass of indefinite chemical composition and containing up to 15% cobalt oxide in a 70% silicaceous matrix. Again, although a common practice at contemporary English porcelain manufactories in the eighteenth century, there is no record of these minor additives ever having been used at Lambeth by Eleanor Coade according to her accepted recipe that was revealed later (Fowler 1850).

5.2 Mineral Formation in Artificial Stone Paste Identified at the Kiln Temperatures

The first thing that comes to the attention of researchers of ceramic manufacturing is that the chemical composition of the raw materials used in the paste formulations for earthenwares, stonewares and porcelains bears little resemblance to the final mineralogical composition of the fired ceramic. This is only to be expected because of the complexity of the chemical reactions that occur at the high temperatures achieved in the kiln to which the paste will be subjected and the final mineral composition will be dependent upon several factors: these include the length of time taken to achieve the final kiln temperature (the heating phase) and the time that the piece being fired is being maintained at the final elevated temperature. The composition of the micro domains that comprise the heterogeneous mixture of the ceramic paste being held at that temperature depends primarily upon the effectiveness of the original grinding efficiency of the component raw materials—basically, the finer the ground frit used in the original paste manufacture then the better the resulting homogeneity that ensues for the chemical reactions to take place at the high kiln temperatures. Finally, the cooling phase is also important, whereby the fired ceramic is then brought down form kiln temperature to ambient temperature. In most thermal chemical reactions conducted in the laboratory an essential component of the procedure is stirring, which enables fresh materials to come together to react and in cases where the efficient stirring or agitation of the reaction mixture is not possible then regions of the reaction mixture can undergo specific and localised reactions to form unusual products where one or more components may then be in an artificial excess. Such is found in cases of living polymerisation reactants in viscous solutions, for example, where the increased viscosity of the reaction mixture prevents the homogeneity of mixing that would be otherwise be preferred to produce the desired reaction product: particularly troublesome in this case is the effect of the reaction mixture being in contact with the material walls of the vessel containing it or the stirrer blades and undergoing unpredictable side reactions which eventually will cause pollution of the main reaction mixture. The stirring speed is also critical in such experiments to continually expose fresh materials to come into contact with each other for reaction to take place. This is not unlike the conditions which are experienced in a ceramic firing kiln, which are exacerbated further by the necessary inability to stir the paste mixture which has now been compounded into a protype artefact, when really the fineness of grinding and the efficiency of the original paste mixing that took place in the manufactory pre-paste preparation now actually dictate the availability of the raw materials to come together effectively and react.

There is no doubt that the empiricism engendered by the trials and experiments that were undertaken by the manufactory proprietors did work well practically for their ceramic syntheses despite their lack of precise knowledge about what chemical reactions were taking place at the high kiln temperatures being operated (Shaw, *The Chemistry of the Several Natural and Artificial Heterogeneous Compounds Used in Manufacturing Porcelain, Glass and Pottery*, Simeon Shaw Personal Publishing, London, 1837). These are now well understood in terms of the ternary elemental oxide diagrams constructed by technologists and solid state ceramic scientists which have identified the critical *liquidus* regions which have to be attained with precise compositions and kiln temperature controls, both of which were generally outside the required precision operating in the eighteenth and early nineteenth centuries. As an example, the ternary elemental oxide diagram drawn by Professor Victor Owen in his studies of Derby porcelain (Owen and Barkla 1997) graphically provides information about the interrelationship of key minerals in the raw materials from which some important deductions can be made which reflect on the potentially heavy kiln losses that are evident from the fragments of sagged porcelain, the presence of firing cracks or overfired shards: Fig. 5.2 shows a typical ternary diagram for contemporary Derby porcelain, adapted from Owen and Barkla (1997)—the main eutectic mineral is anorthite, $CaAl_2Si_2O_8$ and the mullite signature is represented as Al_2SiO_5. These ternary diagrams, which were not available to proprietors at the time, reflect the experience of William Duesbury in the late eighteenth century when William Billingsley was apprenticed at the Derby China Works and then latterly chief decorator at Derby and also later by William Billingsley and Samuel Walker between 1817 and 1820, which eventually contributed to the closure and bankruptcy of their Nantgarw China Works in 1820 (Owen et al. 1998; Owen and Morrison 1999). The crux of the matter is centred on the kiln temperature control to preserve the *liquidus* region in the ternary phase diagram, outside of which the prototype paste is incompletely fused and under-fired, or alternatively over-fired and sagged, in which the artefacts lose their integrity of moulding and thereby become misshapen (Edwrads, *Porcelain to Silica Bricks: The Extreme Ceramics of William Weston Young, 1776–1847,* 2019). Kiln control is absolutely critical for high-temperature fired porcelains and stonewares: not only must the high temperature be achieved and maintained within the specified narrow limits for approximately four days, but the temperature ramp (measured in degrees C per minute) in the heating and cooling regimes over a two day period for each process is also critical—heating and cooling regimes therefore require a change in temperature approaching some 25 °C per hour to be achieved during these processes.

The positioning of the artefacts to be fired in the kiln is also critically important: even with the centuries-long kiln firing experience of the Chinese potters, it was recognised that porcelain qualities could be affected by compositional and kiln positional changes. The Emperor Kangxi recorded that the finest quality ceramics made at Jingdezhen were manufactured in a compositional ratio of kaolin:petuntse of 1:1, whereas the less fine porcelains adopted ratios of kaolin:petuntse of 1:2 and 1:3. It can also be appreciated that with a typical dragon kiln charge of 25,000 pieces at Jingdezhen then the range of temperatures experienced within different parts of the kiln would be a critical factor in the final product quality outcome. Despite the Chinese experience of porcelain manufacture, contemporary visitors to the Imperial porcelain manufactory at Jingdezhen noted the large piles of waste fragments and damaged porcelain at the site ascribed to kiln wastage. Eleanor Coade is recorded (Kelly, *Mrs Coade's Stone*, 1990) as commenting that she had to maintain very close observation of her kiln manager during his duties at the kiln as this was a critical part of her artificial stone manufacturing operation and she had to chide him for deserting his post to seek refreshment in local hostelries when he should have been constantly monitoring the fuel feed to the kiln and its operating temperature! Attention was also given to the correct placement of the artificial Coade stone artefacts in the kiln to ensure that they had the most effective exposure to the temperature gradient, especially for the larger objects and this is manifest in the cutaway sectional print of the Coade stone kiln in operation with the *River God, Father Thames*, occupying a central location and surrounded at the edges by paterae, a frieze, keystones, an urn, a plinth and the *Four Seasons* statues (see the Table containing a Digest of the Coade & Sealy Catalogue, which lists these artefacts among the 800 or so offered for sale). This print (shown earlier in Fig. 2.4), taken from the *European Magazine* in 1786, shows the kiln used for the firing of Coade stone, which also possessed the Latin phrase above its entrance "…*Nec edax abolere vetusta* …". Which translates as "Devouring but not Destroying Antiquity". It is recorded that Coade & Sealy were using three kilns in their manufacturing operation, which can be appreciated when it is realised that each kiln would be occupied for at least one week for each ceramic charge. What is especially interesting and relevant to our discussion here is that there is no mention of a wastage occurring in Eleanor Coade's process: this is very surprising as all other contemporary manufactories of stonewares and porcelains do cite their kiln wastage as a problem, attributed to the ineffective kiln temperature control and firing problems that we have highlighted earlier. The Coade & Sealy factory (shown in the print of 1800 in Fig. 5.1) is now buried beneath the site that was cleared for the 1951 Festival of Britain Exhibition, now occupied by the Royal Festival Hall and Jubilee Gardens, and all that remains to mark the manufactory site is a large granite grinding stone, located under a pier of Westminster Bridge, that was used to pulverise the raw materials for artificial stone prior to their incorporation into the paste used for moulding the artefacts (Roberts and Godfrey, *Survey of London*, 1951).

Fig. 5.1 Print by G. Shepherd, *ca.* 1800, of the Coade stone manufactory at King's Arms Stairs, Lambeth, showing centrally a large statue of the River God Father Thames, with a coat-of-arms on the left a large classical column, a bust on a plinth and two urns with several employees engaged in the business of mixing paste for the moulds

5.3 Analytical Techniques Used to Interrogate Ceramics

In the analytical scientific characterisation of ceramics, the major techniques that have been used thus far have been the key quantitative elemental oxide determinations for the major components (SEM/EDAXS) plus the detection of trace elements (XRF) on the one hand and the qualitative identification of the mineralogy (XRD and Raman spectroscopy) on the other (Calpasaro et al. 2019; Caggiani et al. 2021). It would seem at first sight that the analytical data which can be derived from these techniques is mutually incompatible as they address the different themes of elemental versus molecular characterisation but, nevertheless, the separate information from both types of technique can be correlated synergistically to give an interpretation of the practices that were operating in ceramics manufactories, with sometimes surprising results emerging. The first point to be recognised is that the raw materials generally do not each give precise individual signatures that can be assimilated directly into a quantitative comparison with the published information regarding the composition of the recipe or formulation used in the manufactory, For example, the percentage of silica that is indicated from an SEM/EDAXS quantitative elemental oxide measurement does not correlate solely with the amount of silica sand that was

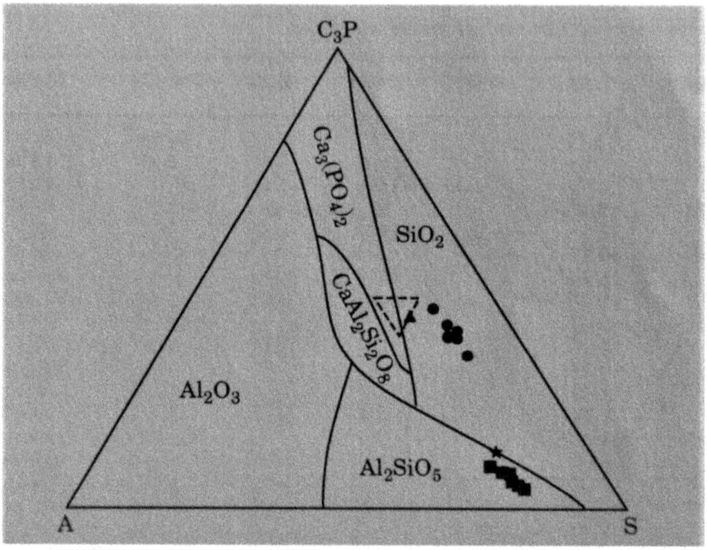

Fig. 5.2 Ternary diagram of Derby porcelain contemporary with Coade stone, *ca.* 1780, showing the interaction of the three major raw materials present in the mixture, namely whitlockite, alumina and silica in the fired ceramic composite at a kiln temperature approaching 1250 °C. The major mineral component at this temperature is anorthite, $CaAl_2Si_2O_8$. Adapted from Owen and Barkla (1997) from St Pierre (1954, 1955). Coade stone would not have a whitlockite, $Ca_3(PO_4)_2$, component as Eleanor Coade did not use calcined bone ash in her recipe

added in the recipe: contributions to this analytical parameter will be afforded from the sand, flint, feldspar, steatite and glass frit (cullet) components, all of which chemically contain the entity SiO_2. Likewise, the elemental oxide calcia, CaO, is found in several raw material components including lime, glass and calcined bone ash (tricalcium phosphate, $Ca_3(PO_4)_2$). However, some elemental oxides do have a particular and unique association with identifiable raw material components, such as phosphorus pentoxide, P_2O_5, with calcined bone ash, PbO with flint glass cullet, and MgO with steatite, $Mg_3(Si_4O_{10})(OH)_2$. The amounts of potash and soda, K_2O and Na_2O, also reflect several contributions from raw material components and not only from their alkaline ash component used, although the ratios of the potash to soda components have occasionally been interpreted analytically in terms of the preferred use of the calcination of terrestrial *versus* marine vegetative sources which are rich in potassium or sodium, respectively. A higher incidence of calcia and soda can also be indicative of a crown glass component being present in the form of cullet. The mineral formation that occurs at the high temperatures adopted in the firing kiln is a function of the chemical changes that have been incurred in the raw materials during the heating cycle, this being typically 40 h at a heating rate of about 5 °C every 10 min, the maintenance of the steady state temperature for a stoneware, artificial stone, hard paste or a soft paste porcelain being achieved between about 1200 and 1400 °C for another 80 h, and finally a steady cooling cycle over a similar period in which the hot vitreous ceramic is annealed to remove stresses and stabilise it for use in the household. In the next section, therefore, it shall be appropriate to explore the mineral formation that can occur chemically during these temperature cycles. Although the changes in the major components, namely silica sand, flints and clays are considered primarily here, the associated thermal changes in other potentially important components are also stated in italic script. This ensures that we do not just consider the chemical changes that take place in the clays and silicaceous materials but recognise that the presence of other raw materials that are potentially present will also affect the final mineralogy, such as the presence of calcium and magnesium ions from lime, calcined bone ash or steatite in an alkaline environment at high temperatures and the reactions they can undergo with the silicates that are formed at elevated temperatures.

5.4 Chemical Reactions That Occur in Raw Material Component Mixtures at Elevated Temperatures

To provide an indication of the complexity of the high temperature chemical reactions that occur in the raw material mixtures which have been compounded into ceramic pastes and fired in the kiln, it will be appropriate to consider sequentially the reactions that can occur at several stages in achieving the final kiln operating temperature. The major changes first occur in kaolin clay, which can be found in several locations in a highly pure state (such as the Cherokee lands in South Carolina, where the white *unaker* clay used in their pottery is estimated to be approximately 97% pure kaolinite), but kaolin is normally itself a mixture and can contain small percentages of related minerals such as muscovite, quartz, anatase, feldspar, illite and montmorillonite. The Dorset ball clay component used by Eleanor Coade for her artificial stone composite contained about 50% kaolinite (derived from *kaolin*, after the village where it was first found in China, Gaoling, near Jingdezhen in Jiangxi province) with the remainder being mica and quartz in approximately equal proportions, but her use of grog would also have included a more complex Dorset ball clay component of about 65% along with approximately equal amounts of quartz sand and feldspar. Firstly, in the heating process of the artificial stone artefacts in the kiln the kaolinite undergoes the following transformations:

100–150 °C Free water is driven off the kaolinite, which can be represented in its commonly used ceramic formulation of $Al_2O_3.2SiO_2.2H_2O$ or mineralogically as $Al_2Si_2O_5(OH)_4$. Both of these formulae give some information about the state of the water held in kaolinite, namely, that some is rather loosely bound as H_2O of coordination, whereas some is more strongly bound chemically as hydroxyl groups (OH) attached to the silica or alumina. Any additional moisture retention in the kaolin clay that has been acquired from its storage and absorption in damp conditions in the manufactory and during its transportation from the mine in Dorset or from the grog would also be eliminated in this low temperature range. *Likewise, the* *preparation of lime from limestone involves a thermal dehydration at 800–900 °C, but again fresh lime will absorb water from the atmosphere to form slaked lime, when hydroxylation will occur to form $Ca(OH)_2$, and then further absorption of carbon dioxide will occur unless it is used immediately. Although Eleanor Coade did not specifically employ the use of lime as a normal additive there is a suggestion that she may have occasionally coated her artefacts with a skim of quicklime and limestone powder applied as a slip.*

150–200 °C The release of water from the inner pores in the kaolinite phyllosilicate crystal structure occurs, which comprises a tetrahedral sheet of SiO_4 atoms linked to an octahedral sheet of AlO_6 atoms through silicon-oxygen bonds: these sheets can absorb free water and thereby create a material which is a plastic clay for shaping and moulding into pottery. This inner water is now released in this higher temperature range.

200–450 °C Further loss of water occurs in a stage that has been termed a "pre-dehydroxylation", wherein some chemical reorganisation structurally occurring in the kaolinite platelets of the octahedral layer into a tetrahedral layer now commences—this is effectively a precursor stage to the formation of meta-kaolinite. *Water retention by slaked lime conversion of the lime raw material additive, technically represented only by CaO for pure and freshly prepared lime but really involving $Ca(OH)_2$ and potentially $CaCO_3$ as well from the atmospheric absorption of water and carbon dioxide, will now be reversed here.* Also, flint, which is comprised of the minerals chalcedony and moganite, both of which are chemically formulated as SiO_2, undergoes a dehydroxylation of the Si-OH interlayer groups present to form Si-O-Si chains and these then cross-link with the elimination of water in this temperature range.

300–500 °C Organic carbonaceous matter and impurities in the kaolinite clay are burned off: some observers have noted that the clay appearance thereafter is noticeably whiter in colour. *Similarly, this would occur for any lime additive component.*

450–600 °C An endothermic reaction now takes place after water loss has occurred in which meta-kaolinite is formed from a purely internal structural rearrangement of the silicate layers. The time needed for this to take place in the kiln has been estimated at about 90 min, with a resultant quantitative degree of dehydroxylation of $D_{dehydr} = 0.98$. This process is generally represented by the following chemical equation, which although correct chemically does disguise the earlier dehydration and dehydroxylation stages that have already occurred:

$$\underset{\text{Kaolinite}}{Al_2Si_2O_5(OH)_4} \rightarrow \underset{\text{Meta-kaolinite}}{Al_2Si_2O_7} + 2H_2O$$

This reaction results in the internal fracturing of the original silicate structures and the creation of new host silicaceous lattices. According to this equation, all hydroxyl groups have now been removed from the specimens.

600–800 °C Further increase in temperature now produces an increase in the *pozzolanic* activity and reactivity of the meta-kaolinite, which describes its ability to react with any calcium hydroxide or oxide present in quicklime to form products with cementitious properties.

830–870 °C *Calcite $CaCO_3$ and dolomite $CaMg(CO_3)_2$ decompose to calcia and magnesia, CaO and MgO, eliminating carbon dioxide, which can be problematic for the prototype ceramics being fired as the gas bubbles are released into a viscous, vitrifying siliceous matrix, thereby causing holes and fissures which can destroy the resulting ceramic texture, translucency and surface appearance.*

$$CaCO_3 \rightarrow CaO + CO_2$$
$$CaMg(CO_3)_2 \rightarrow CaO + MgO + 2CO_2$$

Also in this temperature range, CaO reacts with silica to form wollastonite, $CaSiO_3$. The calcia, if not derived from a quicklime additive also occurs in the glass cullet used as a component and promotes this reaction.

925–950 °C An internal structural rearrangement in the meta-kaolinite structure now occurs with the elimination of some silica molecules to form an aluminium-silicon spinel structure, sometimes referred to as a *gamma*-alumina structure.

$$2Al_2Si_2O_7 \rightarrow Al_4Si_3O_{12} + SiO_2$$

In this temperature range diopside is formed from the reaction of the calcium and any magnesium ions present with the silica that has been released which is usually written as:

$$2SiO_2 + CaMg(CO_3)_2 \rightarrow CaMgSi_2O_6 + 2CO_2$$

Although, of course, by the time this temperature has been attained any dolomite has been itself converted into the oxides CaO and MgO and the carbon dioxide has already been released, so a better descriptive equation would be written as:

$$2SiO_2 + CaO + MgO \rightarrow CaMgSi_2O_6$$

Furthermore, in the presence of Fe^{2+}, as found in impure river sand, magnesosilicates and ferrosilicates can be formed in this temperature range, such as enstatite $MgSiO_3$ and forsterite Mg_2SiO_4, fayalite Fe_2SiO_4 and ferrosilite $FeSiO_3$. Mixed silicates in the olivine series, $Mg_{2-x}Fe_xSiO4$, are also known to occur with the end members being forsterite and fayalite. Above 900 °C, diffusion occurs of elements between the body surface and its substrate and in the reverse direction silica diffuses from the body into the surface. At this temperature, sanidine, $KAlSi_3O_8$, the high temperature stable plagioclase, can form from its more stable room temperature sister feldspar polymorphs, microcline and orthoclase, also formulated as $KAlSi_3O_8$. The sodium analogue is albite, $NaAlSi_3O_8$.

1000–1050 °C The structurally rearranged meta-kaolinite now converts to mullite crystals in the form of platelets and also forms the high temperature stable form of silica, cristobalite:

$$\underset{\text{Meta-Kaolinite}}{3Al_4Si_3O_{12}} \rightarrow \underset{\text{Mullite}}{2(3Al_2O_3.2SiO_2)} + \underset{\text{Cristobalite}}{5SiO_2}$$

Between 900 and 1000 °C, the moganite component in flints which has been partially transformed at lower temperatures now also converts to cristobalite, the high temperature form of silica. Also, in the presence of alumina, Al_2O_3, calcia and silica react to form gehlenite, $Ca_2Al(SiAlO_7)$. The presence of mullite is an analytical indicator of a high-temperature kiln firing ceramic process.

1400 °C A structural transformation from platelet mullite to needle like mullite crystals now takes place, increasing the physical strength of the structure. The total mass loss incurred by pure kaolinite during these thermal procedures and reactions from room temperature to the maximum kiln temperature achieved here is estimated experimentally at 14%, which is a significant mass loss, without invoking further reaction schemes that will inevitable occur in the presence of other raw material components. This number correlates well with Eleanor Coade's cast moulds which were designed for a 13% reduction in size of the completed artificial stone paste artefact after firing. Eleanor was very keen to produce artefacts of the correct size specified by her clients (several accounts note a strict size measurement of the final artefact to better than about 1/10th of an inch, approximately 2.5 mm) so this reduction incurred to the paste in the firing process would necessarily have to have been carefully monitored. The relative sizes of component statuary would be critical for their incorporation into coats-of-arms and heraldic devices which may be required to fit into pediments and restricted spaces on funerary monuments (Edwards, *Welsh Armorial Porcelains: Nantgarw and Swansea Crested China*, 2022).

The situation is complicated further especially for many porcelains which have the presence of a phosphatic component in the form of calcined bone ash (St Pierre 1954, 1955), which not only provides an additional source of calcium ions but also phosphate (V) ions which can form complexes at high temperature with silicaceous minerals, such as bytownite, $Ca_xNa_{1-x}[Al(Al,Si)Si_2O_8]$, anorthite $CaAl_2Si2O_8$, albite $NaAlSi_3O_8$ and also form whitlockite, $Ca_9(Mg,Fe)(PO_4)_6 PO_3.OH$. Clearly, the kiln operating temperature plays a major part in the creation of the minerals that are formed in a ceramic mixture at high temperatures: earthenwares (terracottas and slipwares) require a kiln temperature of between 600 and 900 °C, whereas stonewares (which include Cologne ware, ironstone china, Coade stone, Bottger ware and Crouch ware) are fired at the higher temperatures of between 1150 and 1300 °C and porcelains sometimes require even higher kiln temperatures of between 1200 and 1400 °C. Stonewares differ from earthenwares not just in their firing temperatures but compositionally in their component clays and in whether or not sand has been used as a component raw material: the latter is a rather grey area which compromises many attempts made at discrete definitions of a stoneware that have been proposed historically, several of which simply state that stonewares do not contain flints, quartz, mica, feldspar or sand, which is certainly not applicable to several types of stoneware that have been included in this category above—such as Coade stone, ironstone, Crouch ware (which is named after its clay which was sourced alternatively from Crich in Derbyshire or Creussen in Bavaria), Chinese Yixing wares and Cologne wares.

Some manufactory proprietors either consciously or unconsciously added sulfur in the form of a sulfate, usually in the form of an alum such as potash alum, $KAl(SO_4)_2.12H_2O$, or otherwise $K_2SO_4.Al_2(SO_4)_3.24H_2O$, or perhaps as Na_2SO_4, which is a component of naturally occurring alkaline natrons which also contain sodium carbonate and bicarbonate. In this event, the presence of sulfur is recorded analytically, and expressed either as the elemental oxides, sulfur trioxide SO_3 or sulfate SO_4^{2-}, but new sulfur minerals can be formed at elevated temperatures in the kiln, such as antlerite, $Cu_3SO_4(OH)_2$ and aubertite $CuAl(SO_4)_2Cl$, which have both been detected analytically in earthenwares and stonewares where copper salts have also been used as superficial decorative pigments. Alums were certainly added as minor components to several ceramic pastes and recorded by manufactory proprietors but the purpose of this is not clear from a chemical standpoint, although it may be related to the ancient usage of potash alum and its volcanic congener, alunite, $KAl_3(SO_4)_2(OH)_6$, as additives for imparting a vitreous and white lustre to materials in which they were incorporated.

In the discussion of precursors to Eleanor Coade's artificial stone the presence of stonewares is mentioned: these appeared on the scene first in Elizabethan England around 1560 and they most highly prized, especially those imported from Cologne and termed "Cologne wares".

5.5 Key Mineral Identification in High-Temperature Fired Ceramics

The presence or otherwise of certain minerals in fired ceramics can provide some novel information regarding the firing temperatures, the raw materials and their potential sourcing used in their formulation and the presence of minor additives in the paste. A list of minerals that have been found in fired ceramics, including porcelains and stonewares, is given in Table 5.3. This list is very broad as this is necessitated by the presence of some unusual minerals that can occur when small amounts of the minor components or additives have reacted with the major mineral components at elevated temperatures in the kiln. Other minerals have not been featured here because they are only found when reactions have occurred between pigments that contain transition metals such as nickel, copper, manganese and cobalt with arsenic and sulfur-containing minerals from minor additives in the fused ceramic melt, such as rammelsbergite, skutterudite, fairfieldite and bieberite,

Table 5.3 Minerals potentially found in fired artificial stone with their molecular spectroscopic identifiers

Mineral	Chemical formulation	Firing temperature/ Material interpretation	Spectroscopic bands[a]/ cm^{-1}
Anatase	TiO_2	Low/Impurity in kaolinite	144, 197, 399, 515, 640
Rutile	TiO_2	High/Impurity in kaolinite	445, 610
Mullite	$3Al_2O_3.2SiO_2$	High	305, 340, 415, 530, 600, 720, 870, 960
Wollastonite	$CaSiO_3$	Low	337, 408, 636, 969, 1045
Parawollastonite	$CaSiO_3$	Low	412, 635, 824, 971
Diopside	$CaMgSi_2O_6$	High	323, 356, 391, 666, 1012
Forsterite	Mg_2SiO_4	High	824, 856
Fayalite	Fe_2SiO_4	High	815, 838
Cristobalite	SiO_2	High	230, 416
Enstatite	$MgSiO_3$	Medium/High	164, 240, 345, 666, 689, 1014, 1036
Anorthite	$CaAl_2Si_2O_8$	High	197, 286, 485, 505, 744, 914, 977
Albite	$NaAlSi_3O_8$	High	216, 250, 290, 505, 528, 645, 760, 814, 977
Gehlenite	$Ca_2Al(SiAlO_7)$	High	218, 243, 463, 531, 662, 679, 801, 917, 932, 982, 1009
Moganite	SiO_2	High	211, 503
Quartz	SiO_2	Medium high	210, 467, 960
Microcline	$KAlSi_3O_8$	Low	476,514,993
Haematite	Fe_2O_3	Low/high	225, 245, 292, 410, 610
Sanidine	$KAlSi_3O_8$	Medium high	162, 283, 473, 513, 770, 803, 1004
Pseudobrookite	$FeTiO_5$	Medium	217, 339
Montmorillonite	$(NaCa)_{0.3}(AlMg)_2Si_4O_{10}(OH)$	Medium	430, 710, 785, 840, 915, 1110
Muscovite	$KAl_2(AlSi_3O_{10})(OH)_2$	Low/Medium	189, 290, 370, 588, 669
Illite	$K(AlMgFe)_2(SiAl)_4O_{10}(OH)_2$	Low/Medium	205, 271, 391, 465, 638, 709
Monticellite	$CaMgSiO_4$	Low/Medium	255, 410, 815, 852, 900

[a]The spectroscopic band wavenumber signatures cited here are obtained from non-destructive Raman spectral studies of ceramic bodies and thermal interactions with the glaze. A description of the technique is provided in Edwards, Howell G.M., *Porcelain Analysis and Its Role in the Forensic Attribution of Ceramic Specimens,* Springer-Nature, Dordrecht, The Netherlands, 2021

which have all nevertheless been identified spectroscopically in fired and decorated stonewares and majolica, if not directly in undecorated artificial stones.

The first impression that could arise from the data emerging from a body paste and glaze analyses is that there are two different regimes being explored: the quantitative elemental oxide data refer to the raw material components that were used whilst the qualitative molecular spectral data are more pertinent to the reactions that have been undergone by the prototype porcelain, stoneware or earthenware during the firing process of the paste artefacts in the kiln. The former data provide information about the raw materials that were used and also their potential sourcing whereas the latter tell us more about the kiln operations and the temperatures that were achieved, which can then lead to a better interpretation of the type of ceramic that is being considered, which is otherwise often not clearly definable. Of more recent application is the analysis of the pigments from an elemental and molecular basis, which has revealed novel information about their sourcing, preparation and usage, and this can impact significantly on the body paste analyses (Colomban et al. 2020).

References

J.A. Anderson, The Duesbury papers – a personal view of the management, *c.* 1785–1796. Derby Porcelain Int. Soc. J. **II**, 1–8 (1991)

J.A. Anderson, *Derby Porcelain and the Early English Fine Ceramics Industry*. Ph.D. thesis (University of Leicester, 2000)

C.J. Brooke, H.G.M. Edwards, F. Kosek, J. Jehlicka, C. Stanford, M.J. Champion, The analysis of two historic fired artificial stonewares: Coade stone (18th-19th Centuries) and a recent discovery from Oxnead House, Norfolk (16th-17th Centuries). Archaeometry **65**(4), 691–698 (2022)

M.C. Caggiani, G. Barone, L. de Ferri, R. Laviano, A. Mangone, P. Mazzoleni, Raman and SEM-EDS investigations into the technical aspects of mediaeval and renaissance ceramics from Southern Italy. J. Raman Spectrosc. **52**, 186–198 (2021)

E. Calpasaro, M. Jauregui, H. Morillas, G. Arana, J.G. Inanez, Non-destructive screening methodology based on ED-XRF for the classification of mediaeval and post-mediaeval archaeological ceramics. Ceram. Int. **45**, 10672–10683 (2019)

S.A. Church, *English Porcelain: A Handbook to the China Made in England During the Eighteenth Century as Illustrated by Specimens Found Chiefly in the National Collections* (Chapman & Hall Ltd., London, 1894)

P. Colomban, H.G.M. Edwards, C. Fountain, Raman spectroscopic and SEM/EDAXS analysis of highly translucent Nantgarw porcelain. J. Eur. Ceram. Soc. **40**, 4664–4675 (2020)

H.G.M. Edwards, *Porcelain to Silica Bricks: The Extreme Ceramics of William Weston Young (1776–1847)* (Springer-Nature, Dordrecht, 2019)

H.G.M. Edwards, *Porcelain Analysis and Its Role in the Forensic Attribution of Ceramic Specimens* (Springer-Nature, Dordrecht, 2021)

H.G.M. Edwards, *Welsh Armorial Porcelain: Nantgarw and Swansea Crested China*, Preservation of Cultural Heritage Series (Springer-Nature, Cham, 2022)

C. Fowler, Some remarks on terracotta and artificial stone as connected with architecture. Civil Eng. Archit. J. **13**, 215–216 (1850)

I. Freestone, Forgotten but not lost: the secret of Coade stone. Proc. Geol. Assoc. 135–138 (1991)

R.B. Heinemann, M. Maggetti, *Ancient and Historical Ceramics: Materials, Technology, Art and Culinary Traditions* (Schweizerbart, Stuttgart, 2014), pp. 12–36. and 255–278

L.L. Jewitt, *The Ceramic Art of Great Britain from the Prehistoric Times Down to the Present Day*, vol I and II (Virtue & Co. Ltd/Paternoster Row, London, 1878)

D. Jonyunaite, J. Senvaiteiu, A. Beganskieni, A. Kariena, Spectroscopic analysis of blue cobalt smalt pigment. Vib. Spectrosc. **52**, 158–162 (2010)

H. Judah, *Lapidarium: The Secret Lives of Stones* (John Murray Publishing, London, 2022)

A. Kelly, *Mrs Coade's Stone* (Self-Publishing Association, Worcestershire, 1990)

J.V. Owen, Antique porcelain 101: a primer on the chemical analysis and interpretation of eighteenth century British wares, in *Ceramics in America*, ed. by R. Hunter, (The Chipstone Foundation, Milwaukee, 2002), pp. 39–61

J.V. Owen, R.D. Barkla, Compositional characteristics of 18th century Derby porcelains: recipe changes, phase transformations and melt fertility. J. Archaeol. Sci. **24**, 127–140 (1997)

J.V. Owen, M. Morrison, Sagged Phosphatic Nantgarw porcelain (ca. 1813–1820): casualty of overfiring or a fertile paste? Geoarchaeology **14**, 313–332 (1999)

J.V. Owen, J.O. Wilstead, R.W. Williams, T.E. Day, A tale of two cities: compositional characteristics of some Nantgarw and Swansea porcelains and their implications for kiln wastage. J. Archaeol. Sci. **25**, 359–375 (1998)

W.R.H. Ramsay, E.G. Ramsay, A case for the production of the earliest commercial hard paste porcelains in the English -speaking world by Edward Heylyn and Thomas Frye in about 1743. Proc. R. Soc. Vic. **120**, 236–256 (2008)

W.R.H. Ramsay, P. Daniels, E.G. Ramsay, *The Limehouse Porcelain Factory: Its Output, Antecedents and the Influence of the Royal Society of London on the Evolution of English Porcelain Based on Composition and Technology* (Invercargill Press, South Island, 2013)

S.H. Roberts, W.H. Godfrey (eds.), *Survey of London*, Chapter 15, Volume X (South Bank and Vauxhall, Institute of Historical Research, London County Council, Lambeth, 1951), pp. 58–61

S. Shaw, *The Chemistry of the Several Natural and Artificial Heterogeneous Compounds Used in Manufacturing Porcelain, Glass and Pottery* (Simeon Shaw Personal Publishing, London, 1837).; reprinted by Scott, Greenwood & Son, Ludgate Hill, London, 1900

P.D.S. St Pierre, Constitution of bone China; I, high temperature phase equilibrium studies in the system Tricalcium Phosphate -Alumina -Silica. J. Am. Chem. Soc. **37**, 243–258 (1954)

P.D.S. St Pierre, Constitution of bone China; II, reactions in bone China bodies. J. Am. Chem. Soc. **38**, 217–222 (1955)

M.S. Tite, M. Bimson, A technological study of English porcelains. Archaeometry **33**, 3–27 (1991)

Chapter 6
The Analysis of Coade Stone Artefacts

Abstract A comprehensive literature survey on the published analyses of Coade stone and some of its contemporaries is followed by an evaluation of the comparative information that is accessible from the experimental techniques used, namely XRF, XRD, SEM/EDAXS and Raman spectroscopy. Consideration of the information obtained from noninvasive and invasive analyses used for excised specimens. The relationship between the elemental oxide percentages determined from the analyses and the composition of the raw materials reported in the published recipe for Coade stone. Determination of the average elemental oxide composition for several genuine Coade stone dated specimens from different locations in the current work and in the previous literature affords the conclusion that the recipe composition has not changed during the production life of the manufactory. The use of XRF particularly for estimating the sourcing of the raw materials from minor heavy metal element content.

Keywords Coade stone analysis · XRF · XRD · SEM/EDAXS · Raman spectroscopy · Elemental oxide composition of genuine Coade stone

6.1 Foundations

Unlike porcelain and other fired ceramics, where there is now an extensive list of research publications on the chemical analysis of the bodies, glazes and decorative pigments thereon, artificial stone artefacts have received comparatively little attention from the chemical analyst. The first attempt at the scientific study of Chinese porcelains to determine their composition was made by the French polymath and chemist René Antoine Ferchault de Réamur in Paris, but this was in the early eighteenth century, before the firm basis of qualitative and quantitative chemical analysis had been established by Antoine Lavoisier, Sir Humphry Davy (1815) and others (Ferchault de Réamur 1727, 1729, 1739). Early English porcelains that were contemporary with the Coade stone wares, such as Worcester, Derby, Coalport and Caughley, were subjected to destructive wet chemical gravimetric analysis first in the late nineteenth century and exemplified by the classic work of Sir Arthur Church (Church, *English Porcelain: A Handbook to the China Made in England During the*

H. G. M. Edwards, C. J. Brooke, *Coade Stone*, Cultural Heritage Science,
https://doi.org/10.1007/978-3-031-71432-0_6

18th Century as Ilustrated by Specimens Chiefly in the National Collection, 1885 and 1894). These were followed by the wet gravimetric chemical analyses reported by Herbert Eccles and Bernard Rackham (Eccles & Rackham, *Analysis of Specimens of English Porcelain in the Victoria and Albert Museum Collection*, 1922) on specimens from the Lady Schreiber Collection lodged in the Victoria and Albert Museum, South Kensington, London, which have been critically discussed in Edwards (*Nantgarw and Swansea Porcelains; An Analytical Perspective*, 2018). Only relatively recently has chemical analysis been applied to artificial stone specimens from the eighteenth century and earlier—the first analyses being reported in the mid-1980s and early 1990s (Freestone et al. 1985; Freestone 1991): these differ from the earlier wet chemical porcelain analyses in that they have been accomplished using analytical instrumentation, specifically using the SEM/EDAXS and XRD techniques, but they still required sampling specimens that were excised from the artefacts.

Chemical analysis, as its application to the characterisation of porcelains and other ceramics has demonstrated, can determine qualitatively and quantitatively the elemental atomic and /or the molecular mineral composition of fired ceramic materials and a knowledge of their raw material components can then be used for the correct interpretation of the published recipes and formulations through inductive reasoning and deduction. These analytical data not only facilitate the identification of the manufactories concerned but also indicate whether or not any significant changes were actually made by the proprietors to their stated and published, or assumed, recipes during their production period and these then can be quantified and amended accordingly from the initial formulation. It is thus possible to evaluate the constancy of the recipe used and whether or not this was preserved through the lifetime of the factory's production or was amended perhaps with the inclusion of additives or alternative materials, which could reflect difficulties experienced by the proprietors in their sourcing and supply of the components.

In the eighteenth and nineteenth centuries, the proprietors of ceramics manufactories regarded their recipes, formulations and sources of their raw materials as secrets which were generally jealously guarded from their competitors. Attempts were often made to acquire details of these from a rival company's workforce or even by actively encouraging key workers to abscond and work for a rival manufacturer. At King Louis XV's Royal Sevres manufactory in 1745, it was mandated legally that any worker who left their employ without securing first the required permission would be arrested, charged and could face the death penalty (Edwards, *Porcelain Analysis and Its Role in the Forensic Attribution of Ceramic Specimens,* 2022). King Augustus II, The Strong, who was a prime mover and collector of early porcelains and intent upon creating his own porcelain at his Albrechtsburg Castle in Meissen in the early years of the eighteenth century, kept his key workers under heavy guard and restricted their freedom, before he had successfully achieved the objective of making his "white gold" in 1710: even then the later defection of a key worker and kiln manager, Samuel Stoltzel, to Vienna in 1719 resulted in the loss of his recipe and information about the sourcing of his raw materials, which were used effectively to establish the du Paquier factory in Vienna in competition with

Meissen. Generally, in England, ceramics workers were employed by manufactories on an annual contract which was renewable every November at Michaelmas, so encouraging the movement of a skilled workforce between manufactories and, thereby, proprietors consciously restricted sensitive information concerning their recipes, raw material formulations and kiln processes to only a few trusted and loyal employees to minimise its voluntary or involuntary dissemination to competitors. Hence, for the preservation of recipes and formulations of ceramic manufactories only very few loyal people close to the proprietor would normally have access to such critical information that would be of any use to a rival manufacturer and often this would be in the purview of the proprietor, kiln manager and family members only. At the Nantgarw China Works between 1817 and 1820, the proprietor, William Billingsley, would mix the raw materials for his esteemed porcelain paste personally behind locked doors in the cellar of his residence,Tyla Gwyn, on the Nantgarw manufactory site: the key raw material in his porcelain paste mixture that he was keen to preserve secret was calcined bone ash, the inclusion of which in his porcelain pastes Billingsley only shared with his kiln manager Samuel Walker, and he even kept this a closely guarded secret from William Weston Young, his third partner and key financier in the Nantgarw China Works business. When he eventually succeeded in heading up the bankrupt Nantgarw China Works in 1820, after the departure of Billingsley and Walker for the Coalport China Works to work with John Rose, Young tried to resurrect the production of the Nantgarw porcelain paste and the fortunes of the manufactory using residual raw material components on site but he was unable to do this successfully as he did not possess the full details of its porcelain paste and glaze compositions—specifically, he had no idea of the inclusion of the essential key ingredient of finely-ground calcined bone ash which amounted to a significant proportion of the porcelain paste raw material! This calcined bone ash component amounted to between 40% and 45% of the mass of the raw materials used in the porcelain paste at Nantgarw and the presence of this calcium hydroxy-apatite facilitated the production of the minerals anorthite and whitlockite at the high temperatures of the kiln achieved (*ca.* 1300–1400 °C) during the firing process. William Weston Young had thus to invent a new paste and glaze for his experimental porcelain trials at Nantgarw in 1820/1821 but failed to secure client interest for ongoing commissions as he was unaware of the original recipes (Edwards, *Porcelain to Silica Bricks: The Extreme Ceramics of William Weston Young, 1776–1847*, 2019). His porcelains were, therefore, perceived by potential clients to be of an inferior quality to those that were produced originally by Billingsley and Walker at Nantgarw. Similarly, Lewis Weston Dillwyn FRS at the Swansea China Works undertook experimental trials of different porcelain pastes between 1814 and 1817, which he faithfully and meticulously logged in his personal diary and workbook, but he took the added precaution of using a code therein known only to himself: these details have been reproduced from Dillwyn's diary in the works of Eccles & Rackham (*Analysis of Specimens of English Porcelain in the Victoria and Albert Museum Collection,* 1922), Morton Nance (*The Pottery and Porcelain of Swansea and Nantgarw,* 1942) and Edwards (*Nantgarw and Swansea Porcelain: A Scientific Reappraisal*, 2017), where the code has now been deciphered and properly labelled with the actual minerals and materials and their quantities that were involved in these experiments. It is not surprising, therefore, that little detailed information was provided publicly about the actual recipes of the artificial stone being manufactured at the time of Eleanor Coade: even earlier filed patents relating to artificial stone, such as that of John Holt, tend to obfuscate this issue and do not necessarily specify the full details,which could be illegally taken and used by a rival patent-breaking competitor to create similar works and artefacts, one of whom was identified by Holt in a later legal document as "*a pretending architect called Batty Langley*".

The range of analytical instrumentation that has been applied to the characterisation of ceramics generally and artificial stone in particular is rather limited: the most commonly used techniques are XRF and SEM/EDAXS for the elemental determinations and XRD and Raman spectroscopy for the molecular and mineral identification (Edwards, Vandenabeele & Colomban, *Raman Spectroscopy in Cultural Heritage Preservation*, 2023; Borg et al. 2020; Vandenabeele and van Pevenage 2017; Brooke et al. 2022; Freestone 1991). Analytically, therefore, modern methods used to characterise artificial stone artefacts firstly require specimens of indisputably, and preferably dated, genuine products from a particular manufactory to provide standards upon which to base comparisons with unknown or copied versions and specimens. The primary comparators involved then are the elemental oxide percentages and mineral molecular signatures which will, of course, differ with recipe compositions and processing conditions such as the kiln firing temperatures. What is not often readily possible without access to a specified manufacturing recipe is the accurate back-calculation of the amounts of the raw materials involved in the recipe formulation using a process known as analytical induction. The reason for this is that the elemental oxide determinations in the fired ceramic body are generally not specific to a single particular raw material source component: for example, the presence of silica, SiO_2, as determined in the ceramic fired body by SEM/EDAXS or XRF measurements analytically based upon elemental silicon, results from its combined presence in sand, flint, chert, feldspar, clays, glass and talc, which can all appear as source raw materials in many ceramic and stone artefacts, as well as the aptly named "grog", which was a ground waste and previously fired ceramic used as a filler component in artificial stone. So, even when a manufacturer states that they have changed the composition of their paste mixture slightly or may have replaced one component partially with another, a change in the appropriate elemental oxide percentage is seen analytically, but it is not a simple task to correlate that with the specific change in raw material composition that has been undertaken unless one also possess details of the precise recipe formulation and any detailed changes that have been manifest therein. The sourcing of the raw materials used in the preparation of the artificial stone formulation is also highly relevant for the interpretation of the analytical data, particularly for the XRF elemental determinations, where trace elements can contribute significantly to the analytical outcome and to the conclusions made therefrom: for example, in the HHXRF (Hand-Held XRay Fluorescence) studies of Karran and Colston (2016a, b) some 26 elements are determined, most of which occur only in trace quantities (parts per million, ppm: where 1 ppm is 0.0001%), but a detailed analysis of these provides invaluable information about the geological sourcing of the raw materials and, of course, if there has been a change in sourcing of a material adopted by the manufactory during its production schedules.

A burning question that has been raised historically, and which thus far has remained unanswered, is whether or not Eleanor Coade changed Daniel Pincot's recipe upon her acquisition of the manufactory in 1769 and, indeed, if Pincot had reproduced the earlier patented recipe of Richard Holt for his artificial stone in 1722 and used that for his own version of artificial stone in 1767, even though he claimed in the mid-1760s that his version of artificial stone was far superior to that of Holt. There is no doubt that the first patent for artificial stone, but not the first occurrence of this material historically, was made by Richard Holt, of the York Buildings, The Strand and Cuper's Bridge, Lambeth, who was well aware of the activities of unscrupulous competitors who were intent upon discovering by underhand means his recipe and formulation. Although he managed to preserve the secrecy of his artificial stone recipe for some years after the filing of his initial patent, it appears that a certain Batty Langley, by all accounts a rather devious "*bold-faced adventurer of ruthless character*" and a "*pretending architect*" according to Holt, was intent upon acquiring Holt's recipe despite the protection that was afforded by his new patent. In 1731 on the 3rd March, therefore, Holt declared a Deposition through two lawyers, John Mowbray and William Bridgman, in which he recounted the details of his process and actual recipe for his artificial stone in very floral language and occupying some five handwritten pages (Valpy and Kelly 1986). This has been transcribed from Holt's handwritten Deposition and is given in Fig. 6.1 here. The succinct paragraph reproduced below from this Deposition gives the details of Holt's recipe and process and one can but theorise that his revelation of this critical information at this stage was designed as a warning to Langley and others and as a counter to what he perceived was his until then secret recipe being made known unofficially and used elsewhere:

> Take broken Glass, pound sift it very fine, and to cause this Glass to flow still more free, Whereby the substance may yet be more Consolidated: mix with ye Glass a certain proportion of Lead Ore viz. Potter's Ore, unto every five pound of Glass put one pound of Ore. These Substances, mix't with an Equall quantity of Clay, will in strong fire run almost as flued as water, & will soe vitrifye the united particles, as to become one mass, and what will reject and turn off all manner of wett, and will answer all those end's, that are required to bring Clay up to ye duration of Stone by it's vitrification etc. From whence it is observable, that what cannot vitrifye, cannot strengthen ye Substance: Either sand or flint of themselves will never vitrifye, and therefore remain dry and husky in ye Substance: Whilst ye matter's herein proposed are in their nature most easey to melt, thereby to unite strengthen and Consolidate the Clay without Extravagancy of fire, and in its cost most easey and not expensive -The finer ye Glass and the Ore are reduced the greater will be the strength and perfection of ye Thing – and experience will shew, that what is here proposed is an assured Foundation to produce a Substance most easey to Form, and when Formed by Strong fire to be so hardened no Toole can touch it.

The substance of Holt's recipe for his artificial stone, therefore, was 5 lb of broken glass, which was ground very finely with 1 lb of lead oxide, the whole then being mixed with 6 lb of an unspecified clay. Holt advocated against the incorporation of

Transcribed from Richard Holt's original handwritten deposition made to John Mowbray and William Bridgman in 1730 (British Library MS 11394 --- after Valpy 1986.

Dated 3rd March 1731/2.

The design of this Invention of making Artificiall Stone etc Is not only to provide against ye difficulty's and defect's that are in Natural Stone But likewise to bring up clay to a hardnesse and duration at least equall to Stone, and to take ye Impression, and bring into Form the thing proposed farr Quicker and Cheaper than in either Stone or lead can be done.

It is to be observed ye Naturall Stone has no Tye, otherwise than what time gives it in ye Bowells of ye earth, wherein ye grain's of sand, becomeing fine and united hold in a Body together, but not so firm but that rain drove by wind , will force itself into ye pores of Stone, and by degrees moulter and decay it, haveing in itself a quallity of succion and imbibeing water wch occasion's this defect in it.

As to ye property's of Clay, It will take a Form easey, but if the Substance required be thick and weighty, It must have a long time of d rying, and not without (in a manner) falling to pieces, for want of dry stuff to be mixed with it, as Sand Flint etc. and if a sufficient quantity of dry stuff be mixed therewith it will become rotten and crumbling in ye burning when taken out of the Kiln.

The purpose of ye following Composition is not only to tye it's particles faster than in stone is generally found; alsoe to supply Clay with a dry substance, which in whatever quantity it may be needed, shall no way's weaken in burning, but give a Strength thereto, farr superior to itself: By the Nature of this Composition which in strong fire run's free into a most Compact solid Mass, and brings it into a vitrification, and thereby tye's its particles fast, and shoot's off all well or succion by rendring it extreamly hard and impenetrable.

The Nature of this Composition considered will evidence the unavoydable tendency it must of necessity have to the vitrification and strength proposed. The Ingredients, proportion and nature of ye Composition is as follow's viz. Instead of pounded Flint or Sand wch is the dry stuff usually mixed with Clay to prevent its Cracking and enable it to bare fire, but if used in any considerable proportion it will burn dry and weaken ye Substance: Wherefore to have ye advantage of dry stuff without those Inconveniency's -

Take broken Glass, pound sift it very fine, and to cause this Glass to flow still more free, Whereby the substance may yet be more Consolidated : mix with ye Glass a certain proportion of Lead Ore viz. Potter's Ore, unto every five pound of Glass put one pound of Ore. These Substances, mix't with an Equall quantity of Clay, will in strong fire run almost as flued as water, & will soe vitrifye the united particles, as to become one mass, and what will reject and turn off all manner of wett, and will answer all those end's, that are required to bring Clay up to ye duration of Stone by it's vitrification etc. From whence it is observable, that what cannot vitrifye, cannot strengthen ye Substance: Either sand or flint of themselves will never vitrifye, and therefore remain dry and husky in ye Substance: Whilst ye matter's herein proposed are in their nature most easey to melt, thereby to unite strengthen and Consolidate the Clay without Extravagancy of fire, and in its cost most easey and not expensive -The finer ye Glass and the Ore are reduced the greater will be the strength and perfection of ye Thing - and experience will shew, that what is here proposed is an assured Foundation to produce a Substance most easey to Form, and when Formed by Strong fire to be so hardened no Toole can touch it.

But as to that part wch is lyable to Observation and not practicable to conceale wch is ye moulding or printing out , and bringing into form ye Thing desired and then finising and

Fig. 6.1 Transcribed from Richard Holt's original handwritten deposition made to John Mowbray and William Bridgman in 1730 (British Library MS 11394—after Valpy 1986. Dated 3rd March 1731/2. (*Public domain*)

Compleating it. I shall lay down the severall step's necessary to be Observed in that part. admit ye thing desired was a Mercury or ye figure of Fame, the first thing to be done, is to get a draught of ye best representation to be had: from wch ye Modell is to be made in Clay answerable thereto, from wch mould 's must be immediately taken off in plaister. there must be a great care taken in this point, for ye modeller must know what will deliver from a plaister mould, and carefully avoyd all need less undercutt and He whoever formes or takes off ye mould 's from ye Modell, must have experience to know, what number of pieces will be required, and soe to dispose of those pieces, that the Mould shall deliver unhurt, or without breaking aney part off wch shall be formed in ye Mould . after this part is done it must be set to dry in the most easy manner imaginable, when it has settled itself in it's right Form, and before it's got too stiff It must be carefully touched over by a Modeller or Carver, Who with a light touch in ye places wanting, will repair and put life into any parts that may happen to come out imperfect from the Mould: after wch it must be set to dry thoroughly, and then ye severall Limb's and parts to be filled together, and so lay'd by for ye burning.

In the Article of Burning great care must be taken to Contrive a Kiln suitable to ye purpose, and must be so ordered that ye Flame and smoak must be carry'd out without ever comeing at ye Thing that shall be set therein, and will preserve what is to be brought out in a much better Condition. A further practize and Experiences in all ye parts of this Businesse will open new way's, more and more easily to attain to ye perfection of ye things desired, and a little time will make appear That Artificial Stone is capable not only to be made to give a Representation of all thing's done in Naturall Stone, but in quality and duration really to exceed it, by the help's of such assistances that Nature cannot give.

The above is a Foundation of solid Truth and Improvement all that infinite variety of Shapes and Formation it is capable of being putt into, will assuredly affoard matter of as great advantage, as is possible to lay before a Society of Ingeniouse Gentlemen, desirous to Improve Their own money, and does a great service to the Publick.

The above is Mr. Holts Secret communicated to Us - underwritten in Trust and on Honour, which we promise and oblige ourselves in the most solemn manner and by the most binding Oath taken reciprocally among ourselves never to divulge, communicate or make known to any other Person or Person whatsoever , directly or indirectly, without the Privity, Consent, and mutual Advantage of Us underwritten, our Heirs, Executors or Assignees. So help us God. Done this 3d of March 1731/2

Signed : John Mowbray William Bridgman

Fig. 6.1 (continued)

sand and flint, which he said would introduce a degree of instability to the mixture, which became manifest upon an increased friability and detrimental material granulation upon firing. He was also careful to point out that the kiln design must specifically protect the artefacts being fired from a direct heat which could raise problems for the vitrification process, whilst carefully not revealing the critical temperature and times required for this firing process. The key materials analytically registered that would indicate that a Holt recipe was being followed in an artificial stone material, hence, are the presence of a significant quantity of lead oxide (about 8–9%) and the absence of sand and flint, which would have resulted otherwise in a higher silica content in the artefact; some silica and alumina will naturally be registered analytically anyway in Holt's formulation as clay is an aluminium silicate—the primary clay, kaolinite, for example is $Al_2Si_2O_5(OH)_4$. It has been alleged that Richard Holt historically came originally from a potting background in

Liverpool around 1714, where certainly in the late seventeenth and early eighteenth centuries, several potteries were manufacturing earthenwares and fired ceramics, including faience and delftwares, but the more intensive porcelain manufacture that was carried on there with companies such as those of Richard Chaffers, Samuel Gilbody, William Reid, Philip Christian, Richard Champion and Seth Pennington did not commence until much later in the 1750s (Garner & Archer, *English Delftware (Monographs on Pottery and Porcelain)*, 1972). The authors find it surprising that if indeed Holt had come from a ceramics background in Liverpool or elsewhere, that he rejected the use of sand and flint or chert in his ceramic body as these were the staples of earthenware pottery and the later manufacture of high-fired porcelains. It is however, correct to note here that the much-admired Chinese porcelain being imported from Jingdezhen that was then taking Europe by storm only used two components in its manufacture, namely *kaolin* and *petuntse*, the former being an iron-depleted white clay and the latter a variant of feldspar or *china stone*, as was later mined in Cornwall for the English porcelain industry. The absence of sand, flints and glass, therefore, in Chinese porcelain should be noted. However, in the early 1700s, the recipe for Chinese porcelain was unknown in the West until Pere Xavier D'Entrecolles SJ of the Jesuit mission in Peking obtained the details of its formulation in confessional from two of his Chinese artisan converts to Christianity and he then informed his superior in Paris (Pere Orry) of this in two letters of 1717 and 1722, which did not become public knowledge outside France and elsewhere until the early 1730s, following their translation into English (Edwards, *Porcelain Analysis and Its Role in the Forensic Attribution of Ceramic Specimens*, 2022). Although samples of Chinese porcelain and the component raw materials were made available for chemical and mineralogical examination in France by René Ferchault de Réaumur in the 1720s and 1730s (Ferchault de Réamur 1727, 1729, 1739), as noted above the foundations of qualitative and quantitative chemical analysis were not firmly established until later in the century by Antione Lavoisier, Joseph Priestley, Sir Humphry Davy and others. Hence, it is unlikely that Holt would have excluded the siliceous sand and flints from his formulation on account of the Chinese recipe revelation as he had already patented his recipe in 1722 before the details of their recipes became fully known from the Jesuit letters in the English-speaking world. It is interesting that Holt's patent in 1722 actually specified that his initial artificial stone composite did not even include a clay raw material—which most analysts and ceramic historians would now concur would be unworkable ceramically (Holdridge 1956). However, by the time of his later Deposition in 1730 he had obviously revised the composition of his earlier artificial stone recipe significantly, or perhaps had decided to then reveal its correct formulation to include a clay component.

6.2 The First Analysis of Coade Stone Specimens

The first detailed chemical analysis of Coade stone specimens was undertaken by Ian Freestone in 1991 (Freestone 1991), following an earlier preliminary report from the British Museum Research Laboratory (Freestone et al. 1985), and this was reported in the *Proceedings of the Geological Association,* using a verified Coade stone specimen from Belmont House in Lyme Regis, which had been left to Eleanor Coade by her uncle Samuel Coade in 1784 and which she termed her "seaside pavilion". Eleanor Coade had furnished and embellished this building with Coade stone architectural reliefs and a piece of one of these (whose precise source location was not stated in the paper) formed the basis for Professor Freestone's analyses. A very preliminary study had been undertaken and reported hitherto in a more specialist publication (Freestone et al. 1985) and the 1991 paper expanded upon these preliminary results for a wider scientific audience. The analytical data from the SEM/EDAXS study gave a medium-high silica content for Coade stone of 60%, with the other major component being alumina at 30%, a small percentage of iron oxide of 1.4%, as expected from the white clay and fine sand used, and a low calcia content of 1.4%. These data were said to be consistent with the published recipe as given later by Charles Fowler in 1850 (Fowler 1850), a relative of Eleanor Coade, who also confirmed that a number of manufacturers were still using this formulation some ten years or more after the formal closure of the Coade manufactory in 1837. It is known that both Mark Blanchard and John Blashfield were using this Coade recipe and that they were still using the Coade manufactory moulds obtained by purchase at auction after the closure of the Coade business (Blashfield, *A Catalogue of Seven Hundred Articles made in Patent Terra Cotta, and Red and Cane-Coloured Pottery*, 1882; Davis, *Antique Garden Ornaments*, 1998). The published recipe for Coade stone corresponds with the use of a white Dorset ball clay 50%, grog stoneware 20%, Reigate sand 10%, glass 10% and flint 10%. Clearly, the Coade stone recipe is quite distinct from that quoted earlier for Richard Holt's composite artificial stone formulation in that lead oxide is absent and that sand and flint are both present along with the significant clay component. Freestone could confirm that the powdered glass was an integral component in Coade stone to provide a bonding medium and as a flux for the sand and flint at the higher temperatures adopted in the kiln. The SEM/EDAXS experiment showed images of small domains of glass, whose chemical composition approximated to calcium silicate, $CaSiO_3$, reinforcing the suggestion of its role as a calcic fluxing agent for the sand and flint raw material components. In addition, the analytical data were consistent with a firing temperature achieved of around 1150 °C, which was similar to that used earlier by John Dwight in his Fulham Pottery to synthesise his ground-breaking high quality stoneware and porcelains some years before in the 1670s. The glass used was identified as a common bottle glass and not the alternative highly refractory lead glass (or flint glass) as there was a zero signature for a lead oxide elemental oxide percentage (Cable and Smedley 1987). The publication of the recipe for Coade stone and the comment that it was still being manufactured openly in the mid-nineteenth Century was in direct contradiction to the assertion of Thomas (1990) and earlier historians who maintained that Eleanor Coade's recipe was a closely guarded secret that died with her in 1821. What is not clear is how this recipe compared with that used by Daniel Pincot before she took over the manufactory: Eleanor Coade bought Pincot's ailing business in 1769 and through her acute business acumen turned it almost immediately into a highly successful artificial stone manufactory. Historians have suggested that she may have developed and significantly refined Pincot's formulaic recipe for her eponymous artificial stone after her acquisition of the business but the speed with which she turned the business affairs around and the rapidity of her marketing "Coade's artificial stone" upon her acquisition of Pincot's manufactory in 1769 seem to negate this suggestion. This idea certainly does not hold when we compare it with the time lapse required for similar changes that were effected in porcelain bodies carried out by contemporary factory proprietors who had much greater practical experience and more relevant manufacturing ceramic knowledge and behaviour, which Eleanor Coade seemingly did not possess when she took over Pincot's manufactory as her business expertise was in the area of textiles and linens. Clearly, Eleanor's recipe formulation was totally different to that of Richard Holt and we can hypothesise, therefore, that her recipe was probably identical to that of Daniel Pincot, whose business she had bought into in 1769. Another historical *non-sequitur* that has been proposed by several historians was that Eleanor Coade altered Pincot's recipe to include Dorset clay as she was born there: in fact, this is a complete mis-representation of the facts, as Eleanor Coade was born in Exeter, Devon, and her Dorset connection only appears historically when she inherits Belmont House, then known as Burnett's Castle, and which she referred to as her "seaside pavilion", in Lyme Regis in 1784, some 15 years after she had established her Coade stone manufactory, which by that time was up and running very successfully! However, the white secondary ball clays used in the synthesis of her artificial stone were indeed sourced from deposits in Purbeck, around Bovey Tracey and in the Wareham Basin in Dorset in the south-west of England. Sources of kaolinite and primary china clay were also located nearby at St Austell in Cornwall and were highlighted by William Cookworthy in the synthesis of his early English hard paste porcelain at Plymouth, and later at Bristol, in the mid-late eighteenth century and these too were acquired from that source by other prominent porcelain manufacturers.

6.3 Later Coade Stone Analyses

Following the earliest excavation of Eleanor Coade's factory site at Westminster Bridge for its development as the Festival of Britain site in 1951, specimens of raw materials and damaged relics were deposited in the British Museum Research Laboratory (Hamilton 1954) which formed the basis of a later analytical study in 1984 (Freestone et al. 1985), as mentioned above, which determined initially the presence of the secondary ball clay composition and a firing temperature of 1100–1150 °C for Coade stone artefacts. In the mainstream analytical literature only three papers cite the detailed analysis of Coade stone, namely, the work of Freestone (1991; SEM/EDAXS), which enlarged upon the earlier report of Freestone, Bimson and Tite of the British Museum in 1985, Karran and Colston (2016a, b; HHXRF) and Brooke et al. (2022; SEM/EDAXS, XRD, Raman). A fourth paper has appeared in an archaeological report on the analyses of two Coade stone statues on plinths, dated COADE LAMBETH 1793, and was published by English Heritage (Ponting 2003) in which the artificial stone analyses undertaken using laboratory XRF and SEM/EDAXS instrumentation were cited. An interesting comment was made in this latter report that the analytical data from the external surface of one of the specimens demonstrated a significant increase in calcia content compared with that of the bulk specimen taken from the same figure, which was ascribed to a surface treatment of lime wash having being made to the artefact before firing. This important concept has not been recognised analytically elsewhere and needs to be evaluated further: an alternative explanation is that one of the Coade stone figures had been whitewashed with lime after completion—this would be more acceptable analytically than invoking the suggestion of a subsequent firing process during or after manufacture. In comparison with the rather prolific analyses of early porcelains this relative paucity of analytical data is quite remarkable. A summary of these analytical data for Coade stone, which is detailed in Table 6.1, now follows:

- The earliest SEM/EDAXS analyses of Freestone (1991) (one specimen of verified Coade stone from Belmont House, location of the sample not specified), the intermediate analysis of Ponting (2003) comprising two specimens of Coade stone from two figures at West Dean College, and the most recent of Brooke et al. (2022) (three specimens taken from the Belmont House Gatepost, Coade stone paterae) agree well with each other and confirm that the average silica content of Coade Stone is approximately 59%, which therefore represents a medium-high siliceous ceramic body. The alumina content is also high at 30%,

Table 6.1 Analytical data /% elemental oxide for Coade stone and early artificial stone (Oxnead)[a]

Specimen	SiO_2	Al_2O_3	MgO	CaO	Na_2O	K_2O	TiO_2	Fe_2O_3	Reference
Coade	60	30	0.5	1.4	1.3	2.8	1.2	1.4	Freestone (1991)
Coade	57	30.3	0.5	0.8	1.0	3.1	1.1	1.3	Brooke et al. (2022)
Coade[b]	15–35	2.9-5.0	–	–	–	–	–	0.6	Karran and Colston (2016a, b)
Oxnead	39.5	14.3	1.3	14.6	1.0	1.5	0.4	5.3	Brooke et al. (2022)[c]
Fulham Pottery[d]	80	15	0.3	0.4	0.6	3	0.6	0.8	Tite et al. (1986)
West Dean	61.0	31.0	0.7	1.0	0.8	3.1	1.3	1.1	Ponting (2003)

[a]All data were obtained from SEM/EDAXS instrumentation on excised samples except for the entry [b]data which were determined by HHXRF non-invasive portable instrumentation *in situ*

[b]The determination of several of these elements by HHXRF is recognised as being very imprecise if possible at all: for example, elements with atomic numbers less than calcium (20), such as sodium (11), magnesium (12), aluminium (13), silicon (14) and potassium (19) are usually not possible to evaluate with any accuracy, if indeed at all

[c]The Oxnead stone also showed a minor presence of P_2O_5 and Cl^-, at 0.5% and 0.1%, respectively. The high level of iron (III) oxide reflects the use of normal sand in the recipe, rather than of especially fine and white river sand which is low in iron

[d]The Fulham Pottery product is included here for comparison purposes—although not officially described as an artificial stone it nevertheless is a fired clay-based ceramic stoneware, termed a *porcellaneous stoneware*, which was formed into figures and statuary in the late seventeenth century

calcia and magnesia are both low at 1%, soda and potash at 1% and 3% respectively, titania at 1% and the iron oxide content is also low at 1%. The lead oxide signature is not seen at all and it is, therefore, deemed that the inclusion of a flint glass frit was not undertaken, as suspected. Likewise, the absence of a phosphorus pentoxide signature is also noted which correlates with the fact that calcined bone ash was not used as a raw material by Eleanor Coade, unlike many of her contemporaries in the English fine porcelain industry at this time. The added implication is that Eleanor Coade did not utilise grog material sourced from soft paste porcelain or bone china waste fragments as the 10% grog additive in her raw material, would then have contained a significant proportion of whitlockite, $Ca_3(PO_4)_2$, in its composition. These data are consistent with the established formulation recipe for Coade stone that was provided later in the literature by Fowler (1850).

- The average HHXRF data of Karran and Colston (2016a, b) have been included in Table 6.1 for comparison purposes but have required the conversion to elemental oxide percentages and do create a few potential interpretations which require further discussion. In their comprehensive study, an extensive database of some 24 specimens of Coade stone and other later artificial stones (Blashfield, Doulton, Austin & Seeley and Pulham), all comprising statuary and garden ornaments, were analysed, comprising six specimens of Coade stone for the period 1793–1800 from Anglesey Abbey, Cambridgeshire, Croome Park, Worcestershire, and Buscot Park, Oxfordshire, all of which were documented and marked pieces; five specimens of Blashfield (Stamford) stone of the period 1872–1875 from Burghley House, Lincolnshire, and Castle Ashby, Northamptonshire; four specimens of Doulton & Co., Lambeth, stone from the period 1885–1902 from Castle Ashby, Northamptonshire, and one specimen each of cementitious stone from Austin & Seeley (1870) acquired from the lion statue in the Temple Garden, Lincoln, and Pulham stone (1862) from Dunorlan Park, Kent (Hitching, *Rock Landscapes: The Pulham Legacy*, 2012).

  In addition, seven specimens of assumed Coade and Blashfield stone, unmarked, were acquired from various locations for identification and potential differentiation purposes.
- From a later database, further examples of HHXRF data from dated and stamped Coade stone garden statuary were compiled by Karran (2023a, b) which included a Medici Vase (Kedleston Hall, Derbyshire), Borghese Vase (Kedleston Hall, Derbyshire), three Caryatids (Anglesey Abbey, Cambridgeshire), a Druid (Croome Park, Worcestershire), a Sphinx (Croome Park, Worcestershire), two Egyptians (Buscot Park, Oxfordshire), a seated Lion (original location unspecified) and a female figure (original location unspecified). Three further specimens of questionable assignation to Coade stone were a gatepost pier (Belmont House, Dorset), a King George III Bust (formerly located atop the Dunston Pillar and now restored in Lincoln Castle, Lincolnshire) and a Borghese Urn (Stourhead, Wiltshire). The data sets derived for these HHXRF measurements comprising elemental percentages, not elemental oxide percentages, are presented in Table 6.2. Table 6.3 which gives these data converted into elemental oxide percentages is given here.

Table 6.2 HHXRF elemental % analyses of Coade stone artefacts[a]

Artefact	Si	Al	Ca	K	Fe	Pb	Ti	P	S
Medici Vase, Kedleston Hall	30.0	4.1	4.0	0.6	0.6	0.2	0.4	0.4	2.8
Borghese Vase, Kedleston Hall	30.8	3.7	2.6	0.8	0.6	0.2	0.5	0.3	1.9
Caryatid 1, Anglesey Abbey	31.8	4.5	4.3	0.7	0.6	0.2	0.4	0.2	1.2
Caryatid 2, Anglesey Abbey	28.6	5.8	2.5	1.5	0.6	0.1	0.4	0.2	1.4
Caryatid 3, Anglesey Abbey	27.5	5.5	3.7	1.1	0.6	0.2	0.4	0.2	2.1
Druid, Croome Park	28.5	4.3	1.7	1.4	0.6	0.1	0.5	0.1	0.4
Egyptian 1, Buscot Park	28.9	5.3	2.3	1.3	0.6	0.1	0.5	0.1	1.1
Egyptian 2, Buscot Park	28.7	4.8	3.2	1.3	0.6	0.1	0.4	0.1	0.6
Lion	29.6	6.7	1.5	1.4	0.6	0.1	0.5	0.1	0.6
Female figure	29.2	5.0	1.5	1.4	0.6	0.1	0.6	0.2	0.4
Sphinx, Croome Park	30.4	7.1	1.6	1.5	0.6	0.1	0.5	0.1	0.5
Gate Pier, Belmont House	23.0	6.3	4.9	1.3	0.6	0.1	0.5	0.2	0.5[b]
King George III Bust, Lincoln	14.0	3.5	4,6	1.6	1.3	0.1	4.4	0.1	0.3[b]
Urn, Stourhead House	4.3	5.4	0.0	0.4	0.0	0.1	0.0	1.0	24.5[b]
Average Coade Stone/RSD[c]	29 ± 5	4.8 ± 2.3	3.7 ± 1.4	1.2 ± 0.6	0.6 ± 0.1	0.1 ± 0.1	0.5 ± 0.1	0.2 ± 0.2	1.4 ± 0.9
Range /%:	18–36	1.8–6.9	1.2–6.2	0.1–1.5	0.5–0.7	0.1–0.6	0.3–0.6	0.1–1.4	0.4–7.8

[a]Data taken from Karran (2023a, b) unpublished datasets
[b]These three artefacts are clearly not Coade stone
[c]First 11 artefacts only averaged

Table 6.3 HHXRF elemental oxide % analyses of Coade stone artefacts[a]

Artefact	SiO_2	Al_2O_3	CaO	K_2O	Fe_2O_3	PbO	TiO_2	P_2O_5	SO_4^{2-}
Medici Vase, Kedleston Hall	64.2	7.7	5.6	0.7	0.8	0.2	0.7	0.5	8.4
Borghese Vase, Kedleston Hall	63.1	6.3	3.7	1.2	0.8	0.3	0.7	0.3	6.0
Caryatid 1, Anglesey Abbey	68.0	8.5	6.0	0.8	0.8	0.2	0.8	0.3	3.6
Caryatid 2, Anglesey Abbey	61,2	11.0	3.5	1.8	0.8	0.1	0.7	0.3	4.2
Caryatid 3, Anglesey Abbey	58.9	10.4	5.2	1.2	0.8	0.2	0.8	0.3	6.3
Druid, Croome Park	61.0	8.1	2.0	1.7	0.8	0.2	0.8	0.2	1.2
Egyptian 1, Buscot Park	61.9	10.0	1.8	1.6	0.8	0.2	0.8	0.2	1.2
Egyptian 2, Buscot Park	61.4	9.1	1.8	1.6	0.8	0.2	0.7	0.2	1.8
Lion	63.3	12.7	2.0	1.7	0.8	0.2	0.8	0.2	1.8
Female figure	68.0	10.5	1.7	1.7	0.8	0.1	0.8	0.2	1.3
Sphinx, Croome Park	65.0	13.4	2.3	1.8	0.8	0.2	0.8	0.2	1.5
Gate Pier, Belmont House	49.2	11.9	6.9	1.8	0.8	0.2	0.8	0.3	1.5
King George III Bust, Lincoln	32.7	6.8	5.4	1.9	1.3	0.1	0.8	0.2	1.0
Urn, Stourhead House	9.2	2.7	0.0	0.4	0.0	0.1	0.0	1.8	73.4
Average Coade Stone/RSD[b]	63 $\pm$ 4	9.1 $\pm$ 4.0	5.2 $\pm$ 2	1.4 $\pm$ 0.7	0.8 $\pm$ 0.2	0.1 $\pm$ 0.1	0.8 $\pm$ 0.2	0.3 $\pm$ 0.1	4 $\pm$ 3

[a]Data taken from Karran and Colston pre-published datasets and converted to elemental oxide percentages for comparison purposes
[b]First 11 artefacts only averaged

- A separate paper from Karran and Colston (2016a, b) deals with the analysis by HHXRF of a specimen of Coade stone from a garden statue of a Druid in Shugborough Hall, the ancestral home of Admiral Anson and later the Earls of Lichfield (Karran 2023a, b).
- The HHXRF data as presented by Karran and Colston suffers from rather poor precision on several important elemental oxide determinations, especially silica, alumina and calcia, which is a reflection not of their experiments but rather of the analytical technique adopted, namely their use of non-invasive portable HHXRF instrumentation (Table 6.4). Also, the determinations of sodium and magnesium are not offered as these elements are of too small an atomic number to register for satisfactory detection using this portable technique, but in all some 26 elements are detected and form the basis of their detection protocol for the differentiation between Coade, Blashfield and Doulton stone. Karran and Colston attribute the imprecision of their determinations to specimen heterogeneity (based upon

Table 6.4 Analysis of the HHXRF data RSD percentages for Coade stone specimens from Karran and Colston (2016a, b)

Element	Average RSD%[a]	Range	Mean RSD%[b]
Ba	8	15–3	7
Sn	74	224–25	109
Nb	32	78–0	17
Zr	7	9–6	5
Sr	8	10–4	7
Rb	5	12–0	5
As	90	173–0	69
Pb	56	85–23	42
Zn	46	111–21	39
Cu	52	68–24	39
Fe	7	13–4	6
Ni	71	155–23	58
Mn	21	46–15	21
Cr	36	55–21	27
V	26	41–12	28
Ti	10	13–5	9
Ca	31	62–20	28
K	24	35–3	19
Al	27	30–19	26
P	33	46–20	41
Si	14	28–4	14
Cl	83	125–50	50
S	58	101–33	41

The assumed Coade stone specimens comprising the Gate Piers at Belmont House, the Sphinx statue at Croome Park and the George III legs at Lincoln Castle, were all adjudged to be genuine Coade stone. The Urn at Castle Ashby, Nymph at Dunorlan Park, and the Caryatid at Anglesey Abbey were all judged to be of Doulton manufacture, as was the bust of King George III in Lincoln Castle, whereas the Urn at Burghley House was indeterminate and was not compatible with Coade, Doulton or Blashfield manufacture

[a]Taking the Coade stone authenticated specimens only

[b]Mean RSD% for all sets of Coade, Doulton and Blashfield specimens of artificial stone

assumed recipe changes) but this cannot be the complete answer and an alternative credibility might possibly be given to compositional changes and potentially different geological sources for the raw materials, which would certainly contribute to the variation in heavy metal content even at their very low concentration levels detected, namely, less than <0.1% down to the ppm level (1 ppm is equivalent to 0.0001%). Hence, the determinations of silica recorded at between 15–35% and alumina between 2.9–5.0% typically represent this imprecision (Table 6.1). The authors of this study, namely Karran and Colston, do not present the percentages of each elemental oxide for the specimens studied but give a table of relative standard deviations (RSD %) (Table 6.4) to illustrate the deviations and precision in their measured experimental elemental data. They comment that the major difference between the Austin & Seeley sample of a cementitious stone and the ceramic Coade stone is the very high calcia percentage of the former, measuring around 30%, in comparison to a level of an order of magnitude smaller for calcium that is found for the Coade stone specimens. This is expected because of the nature of the cementitious stone manufacture from a recipe involving calcium-rich lime and gypsum raw materials. Because of the unreliability in precision of the analytical determination of the percentage of calcia in their samples using the HHXRF technique, Karran and Colston have, therefore, relied upon the detection of the presence or otherwise of heavy metal impurities in the raw materials, such as bismuth, often at ppm concentrations in the fired ceramic specimens, to determine the attribution of an artificial stone specimen: as pointed out here, this could have implications for the sourcing of the raw materials also and, therefore, does only indirectly become associated with a particular manufactory. In contrast, the direct measurement of the major component percentages in each stone specimen such as silica and alumina is also less than reliable with this technique as advanced by the authors themselves. It should be mentioned here that the adoption of portable *in situ* analytical sampling using HHXRF instrumentation was essential for Karran and Colston's work as the required permission to take excised specimens from their garden statuary artefacts was generally not forthcoming from the owners concerned. This is an important and often mandatory factor that has to be recognised in the application of analytical science to cultural heritage objects.

- The Oxnead early artificial stone specimen data that feature in Table 6.1 (Brooke et al. 2022) definitively show the characteristic differences with the later Coade stone data even though the recipe and formulation of this earlier seventeenth century version of artificial stone are no longer extant. A distinctively lower silica percentage for Oxnead stone of 39.5% versus 57% for Coade stone and a significantly lower alumina percentage of 14.3% versus 30.3% for Coade stone is certainly enough to differentiate between these two versions of artificial stone analytically. Oxnead stone also has a significantly higher calcia percentage at 14.6% relative to the Coade stone specimens at 0.8% and the iron oxide percentage is also higher at 5.3%, which itself can be related to Eleanor Coade's specific and preferential selection of white river sand and white clays, which are both very low in their iron (III) oxide content, for her production process.
- Similarly, in Karran and Colston's analyses, Blashfield stone has a low alumina content of 1.5% and Doulton stone a higher phosphorus content of 0.4%, although the stated imprecision of measurement determinations would naturally compromise these relative values in comparison with each other as their RSD percentages suggest (Table 6.5).
- Karran and Colston have defined ten principal discriminating elements that were identified for their differentiation of selected Coade, Blashfield and Doulton stone

Table 6.5 Comparison of elemental oxide /% in Coade, Blashfield, Doulton, Pulham and Austin & Seeley artefacts (ex- Karran and Colston)

Manufacturer	Period	SiO_2	Al_2O_3	CaO	K_2O	Fe_2O_3	PbO	TiO_2	P_2O_5	SO_4^{2-}
Coade	1771–1840	62.1 ± 10.3	9.1 ± 2.8	4.1 ± 2.1	1.3 ± 0.4	0.8 ± 0.1	0.2 ± 0.1	1.0 ± 1.2	0.2 ± 0.2	4.2 ± 3.6
Blashfield	1851–1875	62.1 ± 10.7	9.5 ± 3.8	2,5 ± 1.7	1.3 ± 0.5	1.1 ± 0.3	0.0	0.9 ± 0.8	0.2 ± 0.2	1.8 ± 1.2
Doulton	1854–1902	55.6 ± 10.7	11.3 ± 3.8	0.2 ± 0.2	1.5 ± 0.6	1.1 ± 0.4	0.2 ± 0.2	0.6 ± 0.5	0.5 ± 0.4	6.0 ± 4.5
Pulham	1850–1872	53.5 ± 10.7	7.6 ± 1.9	4.2 ± 2.1	1.5 ± 0.6	1.2 ± 0.5	0.1 ± 0.1	0.9 ± 0.8	0.2 ± 0.2	2.4 ± 2.1
Austin & Seeley	1872	15.0 ± 8	2.2 ± 1.5	57 ± 10[a]	0.8 ± 0.5	2.0 ± 0.7	0.1 ± 0.2	0.2 ± 0.2	0.1 ± 0.1	2.7 ± 1.6

[a]The identification of the Austin & Seeley artefact as an unfired cementitious stone is confirmed by its extraordinarily high CaO content

specimens, namely, Ba, Zr, Sr, Rb, Fe, Ti, K, Al, Bi and P and of these the Sr/Fe and Rb/Zr elemental ratios were the most useful in determining and discriminating between the Coade, Blashfield and Doulton artificial stone products from the unknown or unmarked specimens. The large statue of King George III at Dunston Pillar, Lincolnshire (see later Case Studies Chapter here), was created in 1810 by Joseph Panzetta in Coade stone to celebrate the King's Golden Jubilee of 50 years on the Throne but suffered damage upon being carelessly dismantled in the mid-twentieth century to ensure the safe passage of low flying aircraft returning to the nearby airfield of RAF Coleby Grange, Kesteven, in wartime—this was a night fighter-bomber base in WWII. The legs of the statue, which were kept separately, analyse as Coade stone but the fragmented bust does not, this having significantly elevated values of Fe and Bi, Mn, Cr and Ti. This indicated that some restoration and reassembly had necessarily been carried out to the bust fragments using a modern fired variant stone composition. This result is not really surprising when it is realised that the head and bust were smashed into more than two dozen pieces and these had to be reassembled using adhesives and presumably a cementitious or ceramic filler to create the restored artefact now mounted in Lincoln Castle gardens. The Burghley House urn, believed to be of a Blashfield stone origin, however, did not analyse as Coade stone and demonstrated higher values of Ba, Zr, K and Al, again with a much reduced Ca content. The Dunorlan estate garden nymph, presumed to be made of Pulham stone, was low in Fe, Al and P and seemed to show some overlap with a Doulton body. The Castle Aubrey urn, in contrast, was confirmed to be of a Doulton artificial stone manufacture.

- The significantly low percentage of alumina reported by Karran and Colston across all of their Coade stone specimens needs some evaluation in comparison with the other techniques employed, and this cannot be a real phenomenon. Clearly, the precision of the determination is questionable and can realistically be ascribed to aluminium being situated at the lower threshold of detection and measurement using the HHXRF portable *in situ* technique. A comparison with the earlier laboratory XRF instrumentation of Ponting (2003) on excised specimens reveal that he was able to determine not only alumina, but also soda and potash effectively, as follows from the data obtained from his two statues: silica (67.2, 61.0%, average 64.0%), alumina (26.5, 31.0%, average 28.3%), soda (0.4, 0.8%, average 0.6%), potash (2.2, 3.1%, average 2.7%) magnesia (0.5, 0.7%. average 0.6%), calcia (0.3, 1.0%, average 0.7%), titania (1.4, 1.3%, average 1.4%) and iron oxide (1.2, 1.1%, average 1.2%). These values compare favourably with the SEM/EDAXS determinations of the elemental oxides by Freestone (1991) given as 60%, 30%, 1.3%, 2.8%, 0.5%, 1.4%, 1.2% and 1.4% for silica, alumina, soda, potash, magnesia, calcia, titania and iron oxide, respectively, as indicated in Table 6.1. The laboratory XRF determinations were accomplished with an Eagle standardless system with a vacuum/helium chamber atmosphere, which gives an improved detection capability for the lighter elements. At face value, therefore, it would be difficult to accept that the materials analysed by Karran & Colston were Coade stone at all from their alumina determinations solely (which were estimated in the range 2.9–5.0% for alumina), and which are significantly lower than the expected range of values determined by SEM/EDAXS, but it must be borne in mind that they were able to extract analytical information non-invasively from specimens that could not be sampled by excision and by the physical removal of material for laboratory analyses.

Secondly, Karran and Colston were clearly able to attribute correctly their three questionable artefacts of Coade stone, namely the gatehouse pier (Belmont House). King George III Bust (Lincoln Castle) and the Borghese Urn (Stourhead) as being non-Coade in composition. The reason for this must lie in the fact that all three have been restored and this, of course, will significantly affect their surface composition as interrogated by an analytical technique. In contrast, it is interesting that the two specimens from the gatehouse pier at Belmont House analysed by Brooke et al. (2022) were still compatible with their assignment to Coade stone.

6.3.1 XRF and HHXRF Analyses

It is appropriate here to mention the difference between XRF and HHXRF analytical instrumentation that has been discussed and compared in the above analyses. An excellent comparative description of the use of HHXRF in porcelain analysis has been offered by Bezur and Casadio (2012) in a Chapter in a book edited by Shugar and Mass entitled *Handheld XRF for Art and Archaeology* (2012), which covers a wide range of applications to bronzes, paintings, ceramics, historical papers, metal alloys, sedimentary materials and obsidians. Bezur and Casadio describe the advantages of HHXRF instruments in facilitating the analysis of ceramics *in situ* and avoiding the excision of specimens from artefacts, especially porcelains, which invariably results in surface damage, however minor, that owners or curators are naturally not inclined to permit, citing the work in this context of Hornig-Sutter (1984) and Stiegelschmitt and Tomandl (1985) carried out on several hundred early Meissen porcelain artefacts. This is also a major disadvantage of the SEM/EDAXS technique, which is restricted by the dimensions of the specimen chamber and thus normally requires a small fragment of a ceramic specimen for analysis unless the artefact itself is also small enough to be accommodated wholly in the evacuable chamber: recently, an example of such a case was the important SEM/EDAXS analysis of an early English Bow porcelain snuffbox dating from about 1744, whose small size enabled it to be mounted directly and wholly in the instrument chamber without the need for specimen sampling (Ramsay et al. 2023), but this is an exception rather than the rule. The major disadvantage of HHXRF using Si-PIN detectors as implied earlier is that it fails to determine the elements of lower atomic number (low-Z elements) because of the argon content in air in the beam path which attenuates the signals from these low-Z elements. This means that for ceramics, the important elements, comprising oxygen, sodium and magnesium are indeterminate and aluminium is on the edge of detection with a Z of 13 unless the sample is very close to the aperture of the instrument to minimise the air path: other light elements such as potassium, silicon and calcium are determinable but have rather large errors involved. Some attempts have been made to address this deficiency by using a portable vacuum pump to remove air from the beam path and the use of silicon drift detection which improves the determination of low-Z elements without the need for a vacuum pump and flushing with helium. The use of such an instrument under laboratory conditions for transportable artefacts is certainly facile but is clearly still not practicable for Coade stone figures in the field and *in situ* unless specimens are obtained at source, although the adoption of a portable instrumental variant has been described. Hence, the work of Karran and Colston, using an HHXRF instrument, is questionable in the accuracy and precision of its determination of the aluminium content of Coade and related artificial stones (Karran and Colston 2016a, b; Karran 2023a, b).

It is appropriate here to briefly consider the concepts of accuracy, precision and relative standard deviation in analytical data presentation: the accuracy of a measurement reflects the values that are obtained of a particular parameter over multiple sampling points and usually these may vary from just one or two up to several hundred in number. Accuracy cannot be estimated from a single measurement, which may be subject to specific irregularities that may not be appreciated by the observer at that point, and the practice of taking replicate analytical sampling points over the artefact's surface or subsurface is designed to obviate this occurrence. The simple arithmetic average of the data for these determinations gives a number with a ± figure that spans the range of experimental values and this represents the accuracy in terms of the standard deviation (recorded as, one sigma), wherein a one sigma standard deviation represents a 90% confidence interval to the analyst that the true value of the observation will lie somewhere within the limits of this range. The relative standard deviation (RSD) expresses this as a percentage of the estimated value of the parameter concerned. On the other hand, the precision is totally unrelated to the accuracy in that it is possible analytically to determine the value of a parameter which is not the true value but which may well be accurate because it has been derived from the consideration of many contributing values and measurements but for which an unassessed contribution to a systematic error may have been involved which distorts the true value from the average mean value. In the case of Karran and Colston's HHXRF measurements of the Al content of Coade stone in Table 6.2., for example, the determination of 4.8 ± 2.3% from 11 exemplars would imply that it is quite an accurate measurement (which itself is based on perhaps as many as a hundred individual sampling points) and that the true value of the Al content must therefore lie within the range of 2.5–7.1% with a confidence of 90%. In fact, the true (precise) value for elemental Al in Coade stone, as seen in Table 6.1., is calculated to be 15.9%, which is significantly different to this and well outside the range determined in the HHXRF measurements cited, deviating from the true value by a factor of 220–640%. As highlighted above, this discrepancy is attributable to the inability of the technique to effectively record the signals from elemental aluminium, which is at the extreme limit of its range of detection.

6.4 Summary of Analytical Work and Future Potential

The complementary analytical techniques of elemental oxide determination and molecular spectroscopic mineral fingerprinting have demonstrated their applicability to the identification of Coade stone and some of its precursors and later imitators. The first comment to be made from the limited number of analyses that have been achieved so far is that the two main elemental oxide techniques, SEM/EDAXS and HHXRF, give rather different information and this is not surprising since artificial stone is composed of mainly silica, alumina and calcia with minor components of soda, potash, iron (III) oxide and magnesia. The major problem with HHXRF as highlighted above is that the technique sensitivity is dependent upon the atomic number of the element concerned (being a proportionality to the square root of the elemental atomic number), so whilst it is excellent for lead, mercury and the heavy transition metals it is rather poor for the lighter elements, generally those with atomic numbers that are significantly less than 20. Hence, several of the important elements for artificial stone,which are Si, Al, Ca, Na, Mg and K are near the edge of their detectability and measurement or cannot be evaluated at all using this technique: this can be noted especially for sodium (atomic number 11), magnesium (atomic number 12), aluminium (atomic number 13) and silicon (atomic number 14). Hence, Karran and Colston (2016a, b) were only able to roughly estimate the percentages of these elements within a broad range and these are effectively then rendered almost indeterminate as precise values. However, their detection and estimation of the heavier minor elements was effective in discriminating between genuine Coade stone and the artificial stone artefacts analysed of Blashfield and Doulton stone, particularly on the basis of the relative values of the elemental ratios of Rb/Zr and Sr/Fe that they contained from their raw materials. Hence, although the individual Rb/Zr to Sr/Fe ratios lay in the range between 0.25 to 0.028, respectively, for Coade stone, i.e. a ratio of Rb/Zr: Sr/Fe of 9.9, the corresponding values for Doulton stone were 0.25–0.008 and for Blashfield stone 0.1–0.008, which gave analogous Rb/Zr: Sr/Fe ratios of 31.25 and 12.5, respectively. This was sufficient to facilitate the identification of several of the unknown specimens as being Coade, Blashfield or Doulton stone, although one specimen from an artificial stone urn at Burghley House did not match any of the standards and could well be an interloper as it is apparently not compatible with any of the Coade, Doulton or Blashfield manufacturing products.

An interesting analysis of the data given in Karran and Colston's paper (2016a, b) is obtained when one examines closely the RSD % for each Coade stone sample and correlates this with the range of RSDs encompassed in the determinations. The data are presented in Table 6.2 and can be summarised as follows:

- The average RSD% of each of the six Coade stone specimens taken generally lie within a much larger range of RSD % values, which in most cases would render them indeterminate on comparison with their own analogues. Several ranges are very large, for example, Sn at 22–224 RSD% with an average of 74 RSD% and a Mean (see below) of 109 RSD%.

- The Mean RSD% quoted by Karran and Colston includes all specimens from all three manufacturers: the reason for citing this parameter is unclear unless it assumes that all three used the same recipe and formulation. Hence, it is not surprising that all the mean RSD% values lie within the ranges specified.
- Nevertheless, the use of elemental ratios such as those highlighted above, namely Rb/Zr and Sr/Fe does seem to produce a discriminatory plot which separates the Coade stone from the Doulton stone and Blashfield stone specimens.
- A similar exercise performed on the Blashfield stone data obtained from five dated and documented specimens gives an analogous pattern to that obtained for their Coade stone analogues, with equally large ranges in RSD % noted. One striking difference between the Coade stone data and the Blashfield stone data in this respect is that the Blashfield stone data report a zero RSD% for Sn compared with a 74 RSD% for the corresponding parameter in the Coade stone specimens, which could reflect the use of a flint glass frit (high lead) and unglazed grog in the Blashfield stone manufacturing process. The Pb content range is similar but at least Pb is found to be present, which correlates with this conclusion.

The presence of such differences between the minor elements in Coade stone, Blashfield stone and Doulton stone can only be due to the different sourcing of the raw materials: here, a note of caution is raised since we should not perhaps assume that each manufactory sourced the same raw materials from identical sources during their manufacturing periods: is it realistic to assume that Eleanor Coade and her successors did not change their preferred sources of clays etc. during the 70 years of their manufacture of artificial stone? In which case, the tale becomes far more complex as we know from historical documentation that John Blashfield used the Coade stone recipe formulation to synthesise the Coade stone composite paste and the same moulds to make the artefacts, so if a change in source of the raw materials is attributed to the different heavy elemental presence, then we might perhaps expect that some Coade stone artefacts would also demonstrate a similar difference to be seen over a range of values.

What is clear is that from Table 6.1 we can conclude that the determination of silica, alumina and calcia by SEM/EDAXS is a direct measure of the standard characteristics of the Coade stone recipe—and, firstly, it is necessary to evaluate this for the artificial stone artefacts of Blashfield and Blanchard, who both used Eleanor Coade's recipe, irrespective of any issues that arise from their different raw materials sourcing. Secondly, having verified the analytical measurements for genuine marked Coade stone, Blashfield stone and Blanchard stone items it would then be possible to see if there is in fact any substance to the claim that all three used the same recipe and formulation for their products. Then, finally, an examination of the minor heavy elemental presence should then confirm if this is truly the definitive means of discrimination between them.

The HHXRF measurements of Karran and Colston (2016a, b) also include one example of a cementitious stone artefact from Austin & Seeley: unfortunately, again, the failure of the technique to register with precision the percentage of calcia effectively negates a discrimination between a high-fired ceramic and cold-cured cementitious material being made in this respect, even though we would expect the cementitious cold cured stone artefact to possess about 30% calcia compared with a corresponding value of 1.4% for the Coade stone. SEM/EDAXS instrumentation should, however, be able to distinguish between these two extremes on the basis of a CaO determination.

What is clearly now required analytically is the acquisition of a variety of verified specimens of artificial stone, preferably marked and dated, but perhaps otherwise registered in the account books of Eleanor Coade or William Croggan, to establish the standard database for each manufactory against which future comparisons can then be made and evaluated; here, the gazeteers of Kelly and van Lemmen will be useful for consultation and for the selection of specimens for analysis (Kelly, *Mrs Coade's Stone*, 1990; van Lemmen, *Architectural Ceramics,* 2002; van Lemmen, *Coade Stone*, 2006).Secondly, it would be interesting to examine degraded genuine Coade stone specimens to furnish the chemical information necessary for evaluating their susceptibility to deterioration in a range of different environments—this would also address the question as to whether or not the susceptibility to visible deterioration was a function of the quality of the stone artefact, i.e. for the cases where the Coade stone had visibly degraded, do we observe a diminution in its quality, occurring perhaps from the operation of a poor firing process in which the high temperature stable minerals have not formed properly? In the manufacture of porcelains, for example, it is recorded that the quality of output was dependent upon the particular kiln firing operations employed and that in some cases a slightly inferior product with body blemishes was achieved, contributing to the so-called "factory seconds". In the Imperial Chinese porcelain manufactory at Jingdezhen, three grades of quality porcelain body were registered: only the prime grade was used for the Emperor's personal use and for export, whereas the other grades were disposed of locally or destroyed in waste heaps noted by travellers at the site.

Then, with such a database and more knowledge of the chemical composition of genuine Coade stone artefacts, an analyst should be able to determine unequivocally the true status of unknown specimens which appear to be made of Coade stone visually, but then the question still arises if they are really original, or perhaps if they are ostensibly good later copies manufactured using the same moulds and with a similar composite recipe?

6.4.1 An Analytical Study of a Range Dated Coade Stone Artefacts

Finally, if a genuine and verified specimen of Daniel Pincot's early artificial stone could be obtained it would be possible for the first time scientifically to determine if the recipe was changed when Eleanor Coade took over control of the business in 1769: if pre- and post-1769 specimens of "Coade stone" are in fact identical analytically then this would confirm that no change was made in the recipe at the change-over of Daniel Pincot's business to Eleanor Coade in 1769, and that the future business success was due entirely to her astute business marketing acumen and not to a dramatic change in raw materials formulation and a new recipe resulting in a completely novel artificial "Coade Stone" composition. A major problem in this approach is that little of Pincot's stone was apparently sold directly to his clients and he, like his other contemporaries in the artificial stone business, was forced to sell their output in job lots through specialist London auction houses, where they were often purchased anonymously or in mixed batches by a range of clients and agents, which makes their positive unequivocal identification now and eventual intended artefact locations difficult to assess. In some records, Pincot is recorded as having quoted a client for particular statue or artefact but as was evident from William Croggon's day books later in the Coade factory, these often did not materialise into firm orders or commissions: Alison Kelly (Kelly, *Mrs Coade's Stone*, 1990) refers to this particular feature several times in her gazeteer of Coade stone artefacts and comments how careful one must be not to assume that because an inquiry has been made of the proprietor about the provision of an item that a full commission was placed thereafter—it is recorded by William Croggon on several occasions too that the client concerned returned the artefact, claiming they had never ordered it in the first place, so mistakes in bookkeeping commissions were seemingly quite prevalent.

Secondly, to answer the question as to whether or not Eleanor Coade and her successors changed the formulation during the 70 years of production at the Lambeth manufactory, a range of dated Coade specimens should ideally be studied perhaps from each decade of production: i.e. 1770, 1780, 1790, 1800, 1810, 1820 and 1830. Compositional changes would be reflected in the SEM/EDAXS data obtained from such an analytical investigation and should be indicative of such a practice having been incurred at the manufactory. The Karran & Colston work does not provide a range of Coade specimens over a sufficiently large date range (their verified specimens were all in the range 1793–1813) and it would be a task to acquire specimens from genuine verified Coade stone artefacts ideally encompassing the whole 70 years production at the manufactory. The later studies of Karran (2023a, b) do include four additional stamped Coade stone artefacts, dated CROGGON LAMBETH 1824 and COADE LAMBETH 1840, both dating from after the decease of Eleanor Coade in 1821 and also the Kedleston Medici and Borghese Vases marked COADE LONDON and COADE LONDON Fecit, which are both dated on documentary evidence to *ca.* 1771. The authors find it remarkable that the 1840 date stamp exists as, hitherto, the latest Coade stone artefact has been found on a Royal warrant dating from the start of the reign of Queen Victoria in 1837. From Alison Kelly's gazeteer in her book (*Mrs Coades's Stone*, 1990) some possibilities for dated artefacts in verified marked Coade stone over this production period that might be selected for future analysis could be as follows:

1770–1780

1771: Medici and Borghese Vases, Kedleston Hall.
1777: Four lion statues on Bridge, Burghley House, Stamford.
1778: Medici vase, Castle Howard, Yorkshire.

1780–1790

1781: Window tracery, St Mary's Church, Tetbury.
1788: Two figures of Charity St Mary's School, Leicester.
1789: Coat-of-arms on Carburton Gate, Clumber Park, Retford.

1790–1800

1790: Coat-of arms, St Peter's College, Oxford.
1790: Coat-of-arms and patera, Beverley Market.
1790: Bacchus keystone, Butcher Road, Beverley Market.
1793: Caryatids, Anglesey Abbey, Cambridgeshire.
1796: Coat-of arms, York Barracks.

1800–1810

1804: Coat-of-arms, Ollerton Lodge, Knutsford.
1804: Monument to William Mason, All Saints' Church, Sheffield.
1810: King George III legs, Lincoln Castle.
1810: Tripod with Sphinxes, Tatton Park.

1810–1820

1813: Coat-of-arms of General Lord Manners, Bloxholm Church.
1813: Remains of two vases, Alton Towers.
1814: Medici vase, Elvaston Castle.
1814: Monument to General Bowes, Beverley Minster.
1818: Lion fountain, coats-of-arms and two hawks for gate piers, Farnley Hall, Yorkshire.
1818: Borghese vase, Barnby Manor, Newark.
1819: Balusters, Burghley House, Stamford.

1820–1830

1827: Medici and Borghese Vases, Coleorton Hall, Ashby-de-la-Zouch.

1830–1840

1832: Plaque of lion on gateway, Northwick Park.
1840: Grecian spheres, Kineton House, Gloucestershire.

There are, of course, many more alternative and dated artefacts that could be selected for this exercise but the overriding factor must be that they have not been restored or conserved as this could compromise the analyses, as has been demonstrated convincingly already in Karran and Colston's work (2016a, b) on their analysis of the bust of King George III in Lincoln Castle, which was shattered on its removal from the Dunston Pillar and restored professionally by Coade & Co. some years ago, and by the later work of Karran (2023a, b) on the Stourhead Vase. The above list, however, does show that it should be theoretically possible to sample a range of dated and documented Coade stone artefacts over the whole production period of the manufactory. A potential problem arises in the adoption of the SEM/EDAXS technique for these analyses to preferentially determine directly the silica, alumina, calcia, soda and magnesia percentages for which excised specimens are required from the Coade stone artefacts as portable non-invasive instrumentation is not currently available, unlike the HHXRF instrumental technique, and owners are perhaps understandably reluctant to permit the taking of samples from their ceramic garden artefacts (L. Karran, personal communication).

6.5 Current Analyses of Verified Dated and Potential Coade Stone Specimens

As an integral analytical part of research for this project we have undertaken a comprehensive study of several selected specimens of Coade stone which have been taken from either dated impressed artefacts or from artefacts that have been documented in invoices directly from Eleanor Coade & Co., from William Croggon's notebook, or for which orders and commissions still exist in several house archival records. These comprise the following specimens:

Wentworth Woodhouse, Yorkshire; Lord Rockingham purchased from Daniel Pincot's sale of artificial stone in 1767 a pair of river gods, six keystones and a bust of Antoninus Pius. He also acquired in the 1780s four large figures of the Seasons, six vases and two sphinxes and in December 1814, four small figures of the Seasons (£105) and two large baskets of fruit and flowers (£42): all of these were later sold as estate effects and are now in undisclosed private collections. However, two statues of "*The Dying Gaul*" in presumed Coade stone still exist in the central hallway in Wentworth Woodhouse. However, despite this being an established Coade stone figure, their actual origin and date of acquisition is still a matter of some conjecture and debate and they cannot be assumed to be of a Coade stone composition on the basis of their location.

Stourhead, Mere, Wiltshire; Henry Hoare purchased a Borghese vase for location in the Temple of Flora in his gardens for £12 14s 6d, delivered in 1770 by Eleanor Coade, from an original order that was placed with Daniel Pincot in 1769 for an initial fee that he quoted of £5 5s 0d. This is a very early potential Coade stone artefact, which some historians believe could perhaps have been of Pincot stone, but the documentary evidence indicates that it was delivered to Stourhead a while after Eleanor Coade had acquired the factory in 1769 so technically it should comprise an early Coade stone.

The Dunston Pillar, Lincolnshire: King George III statue; originally a tall masonry pier with a lantern on top erected by Sir Francis Dashwood in the eighteenth century—the lantern was replaced with a large Coade stone figure atop the pillar of King George III modelled by Joseph Panzetta in 1810 to celebrate the monarch's 50 years of accession to the throne. Because of potential hazards to low flying aircraft in wartime, the pillar was reduced in size and the figure broken up in 1940: the pieces are now in storage in Lincoln Castle, where the King's bust has been restored from fragments and mounted separately on a plinth located in the Castle Garden. The torso and legs still remain in pieces in storage in the Castle archive and the legs and cloak were sampled from these fragments for the present work.

Harlaxton Manor, Grantham, Lincolnshire (ex-Witley Court, Worcestershire); Thomas Foley, 3rd Baron Foley, purchased unspecified Coade stone artefacts for £86 14s 0d in 1811 which were used when Witley Court was extensively redesigned by John Nash in 1810. It became the home of Queen Adelaide (1792–1849), widow of King William IV, after he died in 1837. The estate was acquired by the Earls of Dudley later in the nineteenth century, but following a major fire in 1937 the house, which was uninsured, is now in ruins. Most of the house fittings and furnishings that survived the fire were stripped out and sold off and two Coade stone lions were acquired for Harlaxton Manor, Grantham, Lincolnshire by Mrs. Violet van der Elst. The ruins of Witley Court are now administered by English Heritage since 1972. Originally a hunting lodge of ancient foundation built for John of Gaunt, Harlaxton Manor was purchased by Sir David de Ligne in 1619; deserted by 1780, it was inherited later by Gregory Gregory, who lived nearby in Hungerton Hall, and in 1831 he commissioned Anthony Salvin to rebuild the mansion in a baroque Elizabethan style, using his considerable fortune to acquire many classical items. It became a Jesuit seminary in the early twentieth century and is now an overseas campus owned by the University of Evansville, Illinois. The two Coade stone lions, bearing heraldic escutcheons of the Foley family, are situated flanking a flight of stairs leading to an upper garden from a lower courtyard at the front of the house.

Farnley Hall, Otley, Yorkshire; Walter Ramsden Hawksworth Fawkes purchased in 1818 a lion fountain for 40 gns, two falcons for his gatehouse piers (£29 8s 0d) and two coats-of-arms paterae (£29 8s 9d), all in Coade stone. The coats-of-arms are now nowhere to be seen and the fountain and two falcons are in pieces stored in a barn on the estate. The falcons were subjected to an attempted theft some years ago but, during their removal from the 20-feet high gate piers upon which they were standing, they fell and were shattered. The broken fragments in storage afforded the opportunity here to sample both the inner and exterior parts of their Coade stone composition. The Coade stone fountain was also badly damaged a while ago and is now also in pieces in storage, being sampled here along with the two falcons.

Wolterton Hall, Norfolk: a Palladian mansion built by Thomas Ripley for Horatio Walpole in 1727 and completed in 1742, it became one of the four power houses of the region. Horatio Walpole was brother of Sir Robert Walpole, first Prime Minister of Great Britain and Foreign Secretary: Wolterton became the country seat of the Earls of Orford thereafter until the mid -nineteenth century. A vase believed to be of Coade stone provided a specimen from this location.

*Syon Park, Isleworth, Middlesex:*the Lion Gateway, designed by Robert Adam in 1767–9 on the commission of the Duke of Northumberland, whose London Home is Syon House, built on the site of the mediaeval Sion Abbey. The gateway comprises a classical arch surmounted by the Percy lion and originally held several vases of Coade stone in niches, but these are now missing. Adam had originally commissioned another artificial stone manufacturer, George Davy, to provide items for use in the construction of the gateway but these were of an inferior quality and did not survive a hard winter, so he recoursed to their replacement with Coade stone from Eleanor Coade in 1770/1.

It can be seen that these artificial stone specimens, whose production can generally be firmly dated from Coade & Co. records and house accounts, afford a range of samples which span the artificial stone manufacturing period from immediately pre-Eleanor Coade (Daniel Pincot) (1769) through the first decade of production (1770–1780) and into the final decade of production (1810–1820) prior to the death of Eleanor Coade. In addition, our previous analytical studies (Brooke et al. 2022) have encompassed specimens of artificial stones from several locations which have included Eleanor Coade's gate piers at her summer home in Belmont House, Dorset, which span the period (1780–1790).From the analytical data derived from these specimens it should now be possible to determine the compositions of the earliest, mid-range and later Eleanor Coade stone recipe formulations with the added advantage of being able to compare the earliest Coade stone formulation (ca. 1770) that was used at Eleanor Coade's acquisition of his business in 1769 with potentially the earlier Pincot recipe. The list of acquired Coade stone specimens, their sources and their dates of manufacture is summarised as follows;

WW1–3: Dying Gaul statues, two in number, Wentworth Woodhouse, Yorkshire, could be pre-Coade, possibly George Davy's recipe, *ca.* 1767; possibly cementitious, cold-cured and unfired?

S1–3: Borghese Vase, Stourhead, Mere, Wiltshire, 1770;

KG1–4: Dunston Pillar, King George III damaged statue fragments, cloak and legs, 1810;

HM1: Harlaxton Manor, lion holding Foley coat-of-arms, 1810;

WH: Wolterton Hall, urn. Commissioned by Thomas Ripley, co -patentee of earliest artificial stone recipe with John Holt. Expected no clay and presence of ground glass, but fired.

SH: Syon House,Gateway to Syon Park, *ca.* 1771. Taken from possibly George Davy's initial artificial stone work, hence cementitious, but possible interference from later Coade replacements.

FH1—FH5: Farnley Hall, two heraldic falcons and a lion fountain, both affirmed Coade stone, 1818.

The latest work of Karran (2023a, b) using noninvasive HHXRF instrumentation has concluded that whereas the Kedleston Hall Medici and Borghese Vases are of a genuine Coade stone composition, there is some doubt cast about the composition of the Stourhead Borghese Vase, the differences have been attributed in their work to extensive restoration and conservation procedures.

6.5.1 Analytical Techniques

The techniques used here comprised molecular spectroscopic (Raman) and elemental (SEM./EDAXS), which together give the most comprehensive analytical evaluation of Coade stone and its congeners undertaken to date. The combination of these molecular structural and elemental techniques gives a qualitative and quantitative assessment of the composition of fired Coade stone over the range of specimens selected for analysis and described in detail above. In addition, it is now possible to formulate an average composition of genuine Coade stone over the period of production using some 11 specimens. The elemental oxide compositional data accumulated from analytical SEM/EDAXS studies of the specimens sourced and cited above are listed in Table 6.6, which includes the previous published literature work on Coade stone of Freestone (1991), Ponting (2003) and Brooke et al. (2022). Following this, it was possible to now calculate the average composition of genuine Coade stone using 11 discrete specimens from 4 different locations spanning the date range 1787–1818 and from these a standard error deviation (one sigma) and range could now be determined for the first time. In summary of these analytical data in Tables 6.6 and 6.7 the following interpretations can be proposed:

- *Farnley Hall, Otley*: 5 specimens taken from two falcon gatepost mascots, internal and external, and a lion's head fountain. All were documented as Coade stone and dated to 1818. The average composition is silica 62 $\pm$ 2%, alumina 30 $\pm$ 3%, calcia 1.1 $\pm$ 0.2%, magnesia 0.5 $\pm$ 0.1%, soda 0.1 $\pm$ 0.1%, potash 2.6 $\pm$ 0.2%, titania 1.2 $\pm$ 0.2% and iron oxide 1.3 $\pm$ 0.2%. No lead oxide was detected, hence confirming that soda-lime glass was used exclusively and not flint glass. The composition of the internal and external parts of the stone figures is the same, refuting a suggestion made occasionally that Eleanor Coade had perhaps used an inferior material composition for the internal construction of her statues.
- *The King George III statue, Lincoln Castle:* 4 specimens, internal and external, documented as Coade stone and dated to 1810. The average composition is silica 62.0 $\pm$ 2.4%, alumina 26.8%, calcia 1.2 $\pm$ 0.2%, and iron oxide 1.4 $\pm$ 0.2% for three specimens, namely the cloak exterior and interior, and the leg exterior. The interior specimen from the leg gave a significantly lower silica at 54.7%, alumina at 11.5%, a much higher calcia at 14.3% and a higher iron oxide at 5.4%. Magnesia at 0.8%, soda at 1.3%, potash at 2.9% and titania at 1,3% are as expected for Coade stone. However, significant sulfate at 3.0% and phosphorus pentoxide at 4.8% suggest that there has been possible contamination of the leg fragments in storage. The other interior specimen from the cloak fragments is the same as the exterior Coade stone specimen from the cloak. An archaeological report (Cowan 2008), referring to an earlier unpublished assessment report of Skerrit (2006), also mentions the presence of reinforcing iron rods inside the King George III statue — this addresses the question posed by historians as to whether or not Coade and Co. used strengthening struts or metallic dowels for their larger statues, such as the King George statue, as smaller pieces, for example the two

Table 6.6 Analytical data /% elemental oxide composition for Coade stone and other specimens from SEM/EDAXS measurements

Specimen Date	Sample Code[a]	SiO_2	Al_2O_3	MgO	CaO	Na_2O	K_2O	TiO_2	Fe_2O_3	PbO	SO_3	Reference	Total %
1720?	WH1	1.3	0.2	2.3	95.5	–	–	–	0.3	–	–	This work	100.6
1771[b]	WW1	18.5	8.8	0.8	3.6	3.1	1.3	14.4	–	–	47.0	This work	97.5
	WW2	16.8	11.2	1.4	4.3	3.5	1.4	19.0	1.4	–	45.6	This work	95.2
	WW3	6.5	1.4	2.1	50.7	3.2	–	1.4	1.5	–	35.2	This work	102.0
1771	S1	46.5	16.5	0.4	10.0	2.5	3.0	0.6	0.7	–	18.7	This work	98.9
	S2	49.6	23.5	2.0	20.0	2.1	1.3	0.8	0.7	–	1.1	This work	101.1
	S3	54.0	18.2	1.0	9.0	2.6	2.2	0.8	0.9	–	12.0	This work	100.2
1771	SPG1	58.0	28.4	0.5	1.0	1.4	2.8	0.8	0.9	1.0	–	This work	94.8
1787	BH1	60	30	0.5	1.4	1.3	2.8	1.2	1.4	–	–	Freestone (1991)	97.0
1787	BH2	57.0	30.3	0.5	0.8	1.0	3.1	1.1	1.3	–	–	Brooke et al. (2022)	95.1
1793	WD	61	31	0.7	1.0	0.8	3.1	1.3	1.1	–	–	Ponting (2003)	100.0
1810	KG1	63.9	24.6	0.5	4.7	0.9	2.0	1.4	1.4	–	5.7	This work	99.4
	KG2	65.3	24.9	0.8	3.5	1.3	2.4	1.7	1.2	–	–	This work	101.1
	KG3	68.9	24.8	0.5	2.0	0.8	2.4	1.3	1.2	–	–	This work	101.9
	KG4	54.7	11.5	0.8	14.3	1.3	2.9	1.3	5.4	–	3.0	This work	100.0 (P4.8%)
1811	WC1/HM1	55.2	14.7	0.7	3.5	1.2	2.2	2.8	3.1	11.0	–	This work	94.4
1818	FH1	64.0	26.5	0.5	1.2	0.9	2.4	1.2	1.4	–	–	This work	98.1
	FH2	61.7	32.4	0.2	1.8	0.8	2.8	1.4	1.0	–	–	This work	101.1
	FH3	59.1	32.0	0.6	1.3	0.7	2.9	0.4	1.8	–	–	This work	99.8
	FH4	63.1	29.9	0.6	1.1	0.8	2.7	1.0	1.0	–	–	This work	100.2
	FH5	61.7	30.2	0.5	1.0	0.8	2.3	1.1	1.2	–	–	This work	100.9
?[c]	H1	51.1	18.9	2.7	16.5	–	1.0	1.0	7.0	–	–	This work	98.8
Grog		59.6	29.9	0.8	0.5	1.0	2.6	1.1	1.5	–	–	Freestone (1991)	97.0

[a]Specimen Source—*BH* Belmont House, *FH* Farnley Hall, *WW* Wentworth Woodhouse, *S* Stourhead, *KG* King George statue Dunston Pillar, *WC/HM* Witley Court/Harlaxton Manor, *SPG* Syon House Park Gateway, *WH* Wolterton Hall, *WD* West Dean College

[b]These three samples of artificial stone from Wentworth Woodhouse possibly pre-date Eleanor Coade's acquisition of Daniel Pincot's company in 1769 and therefore suffice to establish a baseline for Pincot's recipe for comparison with the later artefacts from Coade & Co., Lambeth. Alternatively, they could represent later acquisitions in the nineteenth century: one is of a sensibly different composition to the other—acquired at different times?

[c]This specimen, labelled H1, was unearthed in a hedgerow near Harewood House, Harrogate. Yorkshire and comprised a cream coloured fired ceramic moulded architectural pillar; although suspected to be potentially Coade stone by a local historian, its inclusion in this table affords the basis of an analytical comparison to be made with the preceding genuine Coade stone artefacts

Table 6.7 Sample data selection for the determination of the average composition of Coade Stone

Date	Sample Code[a]	SiO_2	Al_2O_3	MgO	CaO[a]	Na_2O	K_2O	TiO_2	Fe_2O_3	PbO	SO_3	P_2O_5	Reference
1787	BH1	60	30	0.5	1.4	1.3	2.8	1.2	1.4	–	–	–	Freestone (1991)
1787	BH2	57.0	30.3	0.5	0.8	1.0	3.1	1.1	1.3	–	–	–	Brooke et al. (2022)
1793	WD	61	31	0.7	1.0	0.8	3.1	1.3	1.1	–	–	–	Ponting (2003)
1810	KG1	63.9	24.6	0.5	4.7	0.9	2.0	1.4	1.4	–	5.7	–	This work
	KG2	65.3	24.9	0.8	3.5	1.3	2.4	1.7	1.2	–	–	–	This work
	KG3	68.9	24.8	0.5	2.0	0.8	2.4	1.3	1.2	–	–	–	This work
	KG4[b]	54.7	11.5	0.8	14.3	1.3	2.9	1.3	5.4	–	3.0	4.8	This work
1818	FH1	64.0	26.5	0.5	1.2	0.9	2.4	1.2	1.4	–	–	–	This work
1818	FH2	61.7	32.4	0.2	1.8	0.8	2.8	1.4	1.0	–	–	–	This work
1818	FH3	59.1	32.0	0.6	1.3	0.7	2.9	0.4	1.8	–	–	–	This work
1818	FH4	63.1	29.9	0.6	1.1	0.8	2.7	1.0	1.0	–	–	–	This work
1818	FH5	61.7	30.2	0.5	1.0	0.8	2.3	1.1	1.2	–	–	–	This work
Average % (11 specimens)		61.2	29.3	0.5	1.2	1.0	2.7	1.3	1.3				
Standard deviation ±		4	3	0.3	0.5	0.4	0.4	0.4	0.4				
Range From:		57	27	0.2	0.8	0.7	2.3	1.0	1.0				
To:		64	32	0.8	1.4	1.4	3.1	1.8	1.8				

[a]The CaO data for the KG specimens are abnormally high in value and indicate some cementitious contamination: these have hence been omitted from the averaging process

[b]The KG4 specimen exhibits a, unusually large cementitious contamination and has therefore been omitted from the analysis

falcons at Farnley Hall discussed above, did not display this feature at all. Alison Kelly (1996) had earlier commented upon the construction of the Lord Hill statue atop its column in Shrewsbury, which was assembled from twelve pieces secured using bronze armatures, compared with the King George statue in Lincoln which was made in seven pieces using ferrous material. It was also noted that details of previous restoration of the bust in the 1960s and 1970s were not available but that several types of calcareous cement and resins used could be detected

- *Belmont House, Lyme Regis, Dorset:* Two specimens from a gatepost plaque, documented, not firmly dated but probably *ca.* 1787 (Brooke et al.,), silica 57.0%, alumina 30.3%, calcia 0.8%,magnesia at 0.5% and iron oxide 1.3%, with soda at 1.0% potash at 3.1% and titania at 1,1%. This Coade stone analysis matches closely that of Freestone (Belmont House) and the Farnley Hall and King George III Coade stone specimens cited above (with the exception of the KG III specimen number 4 noted in Table 6.6).
- *Dean Wood College,* analyses of two figures dated 1793 (Ponting 2003), confirms their attribution to Coade stone with silica at 61%, alumina 31%, magnesia 0.7%, calcia 1.0%, soda 0.8%, potash 3.1%, titania 1.3% and iron oxide 1.1%.
- A mathematical average of the genuine confirmed Coade stone specimens (excluding the contaminated KGIII specimen) over the date range of 1787 to 1818 gives a definitive data set of silica 61.2 $\pm$ 4%, alumina 29.3 $\pm$ 3%, magnesia 0.5 $\pm$ 0.3%, calcia 1.2 $\pm$ 0.5%, soda 1.0 $\pm$ 0.4%, potash 2.7 $\pm$ 0.4%, titania 1.3 $\pm$ 0.4% and iron oxide 1.3 $\pm$ 0.4%. This shows the narrow distribution of elemental oxides in the Coade stone composition used over some 30 years and indicates clearly that there was no major change in its composition made over that period.
- *Harrogate find* (HG1): One specimen, cylindrical pipe fragment. Silica 51.1%, alumina 18.9%, magnesia 2.7%, calcia 16.5%, soda zero, potash 1.0%, titania 1,0% and iron oxide 7.0%., indicative of a cementitious composition, and definitely not of Coade stone. The specimen probably therefore dates to the mid-1850s: the higher iron oxide content should also be noted as this indicates the use of poorer quality sand as a raw material.
- *Harlaxton Manor (ex Witley Court):* One specimen, lion mascot, marked and dated COADE 1811. Silica 55.2%, alumina 14.7%, calcia 3.5%, iron oxide 3.1% and sulfate 11%. Low silica and low alumina content indicates that contamination with gypsum is indicated, either as a surface rendering or perhaps through a later restoration.
- *Syon Park Gateway, Isleworth, Middlesex*: One specimen, dated to ca. 1771. Silica 58.0%, alumina 28.4%, magnesia 0.5%, calcia 1.0%, potash 2.8%, soda 1.4%, titania 0.8% and iron oxide 0.9%, and lead oxide 1.0%. Very similar to Coade stone in composition but with a slightly lower silica content, and the presence of lead indicates that some flint glass frit must have used, perhaps in admixture with crown glass. Otherwise, very similar indeed to Coade stone and matches our definitive average range for genuine Coade stone quoted above. Could this be George Davy's original work or perhaps an early Coade formulation where an immediate shortage of soda-lime glass occasioned the use of some flint glass? Alternatively, clients would often purchase "end-of-day" glass from

local glassworks which would be sold off cheaply and this could well have contained an indefinite composition mixture of crown (soda-lime) and flint (lead) glass if both had been used in glass production at the glass works on that day. Caroline Stanford (2024) points out that the Coade firm carried out significant but unknown repairs to the gateway in 1802 and 1816 so an original Davy formulation is not ruled out—but nor is a later repair by the Coade firm trying to ameliorate deterioration at a difficult site.

- *Wolterton Hall*: One specimen dated to ca. 1720. Pre-Coade in date. Silica 1.3%, alumina 0.2%, magnesia 2.3%, calcia 95.5%, zero soda, potash and titania and with iron oxide at 0.3% and sulfate 1.0%. Definitely this does not match Coade stone in composition as very low silica and alumina indicate little or no clay was used as a raw material. Very high calcia indicates a potential cementitious origin, also with some gypsum present. This could therefore be an early example of Richard Holt's recipe — as the omission of sand and flint from his recipe would certainly account for the low silica reading.Also, where is the lead oxide which Holt specified that he added to his mixture in the form of lead ore? It looks like some contamination or restoration has been undertaken here, too. The results therefore seem inconclusive and must be considered as being potentially unlikely to be Holt's work.
- *Stourhead Vase*: Three specimens taken from an allegedly unrestored area near the satyr's head, were provisionally dated to 1771. Silica 50.0%, alumina 19.4%, magnesia 1.1%, calcia 13.0%, soda 2.4%, potash 2.2%, titania 0.7%, iron oxide 0.8% and sulfate 10.6%. Definitely not within the range of Coade stone. Very high calcia and sulfate indicates a cementitious origin. Perhaps contamination is indicated as it is known that the vase has been heavily restored and potentially a limewash could have been applied, as has happened at the West Dean College statue? A resampling would be advisable before definitive conclusions can be drawn.
- *Wentworth Woodhouse, Rotherham*: Three specimens, two from the left-hand Dying Gaul statue and one specimen from the right-hand Dying Gaul statue, both of which are indefinitely dated. The left -hand Dying Gaul statue: average silica 17.7%, alumina 10.0%, calcia 7.9%, iron oxide 1.6%, and sulfate 46.3%, indicating a highly cementitious composition with high gypsum content. The right -hand Dying Gaul statue is significantly different: silica 6.5%, alumina 1.4%, soda 3.2% calcia 50.7%, iron oxide 1.5%, and sulfate 35.2%, very different to the left-hand statue but still undoubtedly of a cementitious composition with gypsum. This does not match Coade stone or potentially that of George Davy (?) at Syon Park.
- Eleanor Coade did not seem to change her recipe for the limited range of dates studied.
- The effect of restoration or surface treatment of the specific artefacts needs to be further evaluated with more extensive sampling.
- Raman spectroscopy is ideally required to assess the presence or otherwise of minerals that are characteristic of high-temperature firing processes. This will help to clarify the presence of later cold-cured cementitious compositions versus the high-temperature fired Coade stone ceramics.

6.6 Statistical Analysis

In order to ascertain the statistical likelihood of the individual samples belonging to the Coade Stone grouping a standard multivariate analysis was undertaken. The analytical dataset, shown in Fig. 6.2, are data which are well suited to the method of cluster analysis (CA), a technique that aims to divide data into groups such that observations in the same group (also termed a cluster) are more similar to each other than to those in other groups or clusters (McNicholas 2016; Brooke 2020; Yao and Xiang 2024). Organizing data into assemblages in this way allows some internal structures to be revealed which are not evident from simple observation of the information alone and which are otherwise hidden in the overall measurement. In addition, classification using the non-parametric technique of principal components analysis (PCA), a method of reducing a complex data set to a lower dimension to help reveal any hidden, simplified structure, has also been employed (Jolliffe 2002; Vidal et al. 2016; Härdle and Simar 2019).

- The analysis was run using R scripts and the results validated by also running the dataset under Minitab and Matlab; all the analyses yielded identical, or near-identical output. Although outliers are clearly present in the dataset (for example, only one sample contains lead oxide), current best practice is to attempt modelling of the dataset as it is, outliers included, rather than removing any outlying observations and fitting a model on a new, truncated dataset (Finch and Bolin 2024). Therefore all data have been used in the analysis.
- PCA has been used here to transform the set of correlated compositional values into the same number of uncorrelated linear combinations or components which include all of the original variance in the variables. The first component has most of the variance, and the remaining components take less. As the technique retains all the information from the original variables, whilst making the new components completely orthogonal, potentially overlapping variables may be eliminated (Harlow 2023). The PCA includes all the data from the physical analysis (i.e. there are no missing values) which conform to the same unit (approximate percentage composition). The results of the PCA are shown in Fig. 6.2, it may be observed that a tight cluster of samples form, including known and signed Coade stone, with strong positive correlation whilst the remaining samples have weak or negative correlation to this group and are split clearly into Wentworth left hand statue 1 and 2, Wentworth right hand statue, and Wolterton Hall.
- Both agglomerative hierarchical clustering and k-means clustering were tested (Giordani et al. 2020) and gave comparable results; the former method is preferred (objects are collected into clusters by combining the closest elements to larger clusters until all objects reach a single cluster). There is a significant advantage in this approach as objects are not just placed in distinct groups, but rather in a hierarchy of clusters which yields more information about the structure in the dataset and reveals which clusters are similar or dissimilar. Thus, it becomes possible to detect outliers and sub-clusters (Lei et al. 2017). For most analyses using clustering the default distance measure is the Euclidean distance

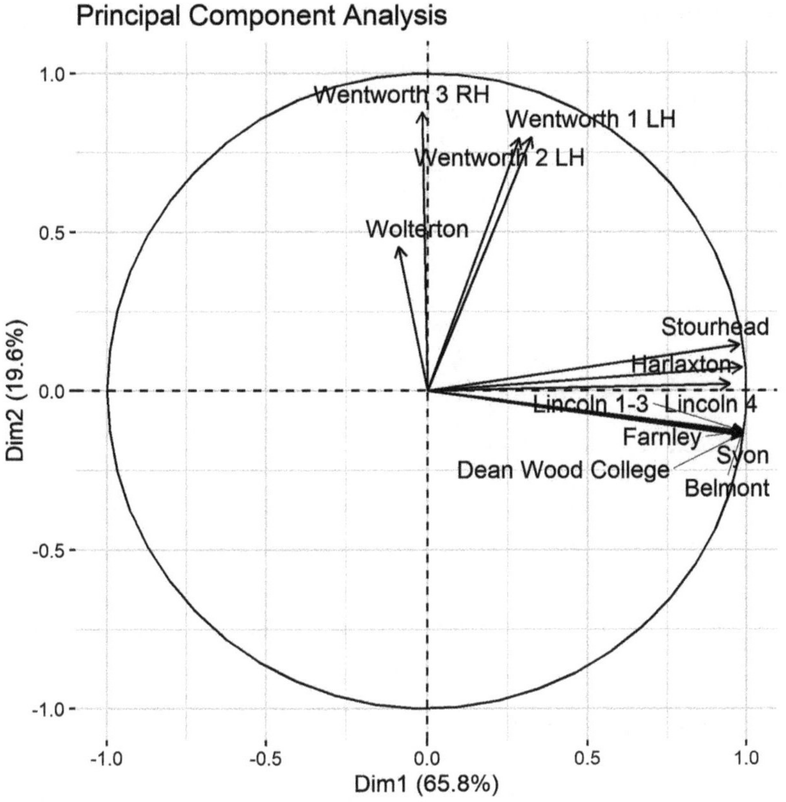

Fig. 6.2 Principal Component Analysis of the sampled analytical dataset

(Aggarwal and Reddy 2014). Complete linkage gives the same dendrogram, regardless of any transformation of the data that preserves the order, and the distance between two clusters is equal to the distance between two farthest objects belonging to different clusters (Wierzchoń and Kłopotek 2018); therefore, in this analysis the Euclidean distance measure and complete linkage have been used. Figure 6.3 shows the dendrogram for the optimum value of four clusters as determined by using the function NbClust in R (Charrad et al. 2014) and yields similar observations as seen in the PCA analysis, except that in the three smaller clusters the Wentworth right hand statue and the Wolterton specimen are grouped together and the contaminated Lincoln KG 4 sample is shown as a separate cluster.

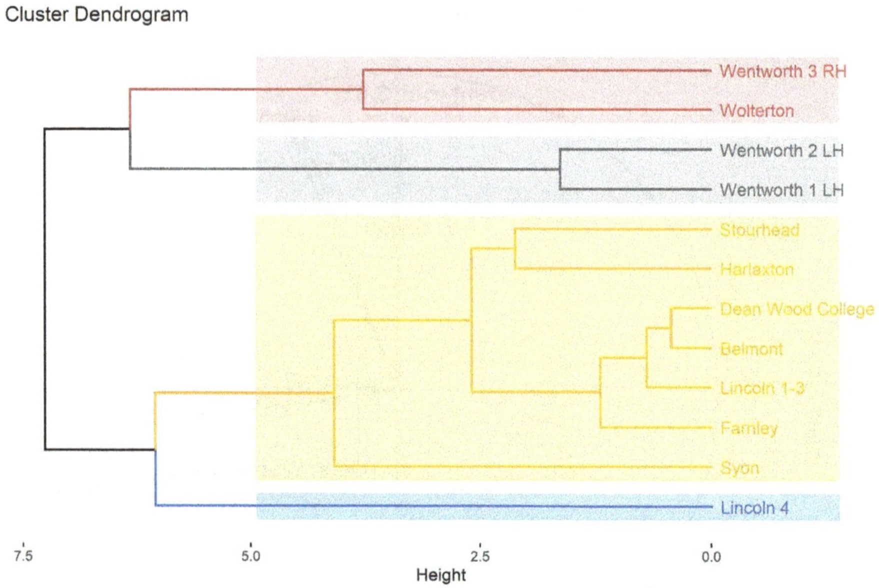

Fig. 6.3 Cluster Analysis of the sampled analytical dataset using Euclidean distance measure and complete linkage

6.7 Raman Spectroscopy

Preliminary studies of the genuine and potential Coade stone specimens have been subjected to laser Raman microspectroscopic analysis using several wavelengths of excitation between 633 and 1064 nm. An example of the type of spectrum obtained at 785 nm in the near infrared is shown in Fig. 6.4 for the cloak fragment of the King George III statue in Lincoln Castle: here, the wavenumber shift range between 100 and 2800 cm^{-1} is shown which is characterised by a broad resonance fluorescence spectrum between 500 and 1300 cm^{-1} which is typical of a high-fired ceramic clay, as expected for Coade stone. Superimposed upon this background are several bands which can be attributed to mineral species, as specified in Table 5.3, such as anatase (an impurity in the clay), cristobalite (the high temperature form of quartz / sand), mullite and wollastonite, which are formed at high temperatures in the kiln. Also, some gypsum is detected from the Raman bands at 1008 and 1133 cm^{-1}, and this can be ascribed to surface contamination from other fragments acquired whilst in storage (Fig. 6.5).

Other specimens of genuine Coade stone from Lincoln Castle, Farnley Hall and Harlaxton Manor give similar spectra, all of which are characterized by the characteristic resonance fluorescence of a fired ceramic upon which are superimposed the vibrational bands of minerals such as anatase, rutile, mullite and wollastonite. The specimens from several other locations, however, such as Wentworth House,

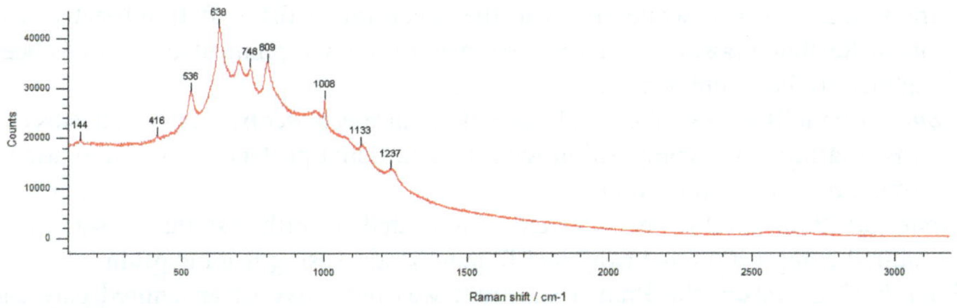

Peak no.	Centre	Height	Width	Area	Absolute inten...	Low edge	High edge
1	536.077	5441.34	20.6343	35533.1	29491	356.729	567.978
2	638.105	12668.8	26.3696	437387	42363.3	569.05	680.445
3	748.337	4479.7	21.5851	214042	34406	681.495	777.122

Fig. 6.4 Raman spectrum of The King George III statue in Lincoln Castle, wavenumber shift range 100–2800 cm^{-1}, laser excitation 785 nm, cloak fragment, showing the evidence for it being a fired ceramic with Raman signals in addition for anatase and gypsum, indicating the presence of contamination through cementitious plaster during storage

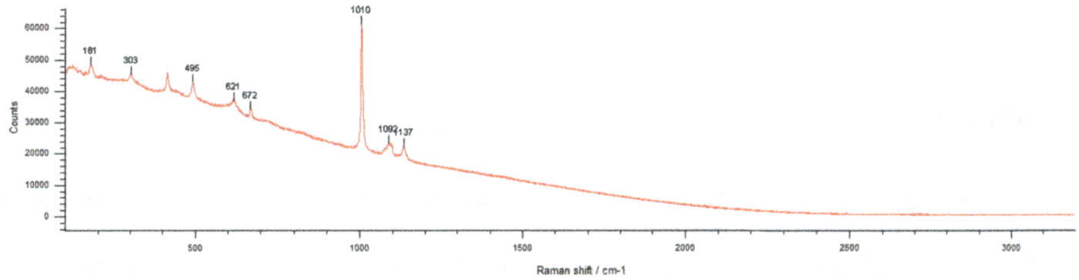

Peak no.	Centre	Height	Width	Area	Absolute inten...	Low edge	High edge
1	180.529	3801.07	11.2921	60860	48735.7	152.922	256.335
2	302.932	2953.07	12.2198	42334.4	45698.7	271.109	380.076
3	495.447	5202.28	9.91661	63499.1	42932.7	468.287	578.641
4	1009.55	40481.9	6.08491	337608	61712.1	973.938	1054.44
5	621.319	-	-	-	37162.4	-	-
6	671.907	-	-	-	34442.1	-	-
7	1091.53	-	-	-	23438.3	-	-
8	1137.42	-	-	-	22618	-	-

Fig. 6.5 Raman spectrum of the Right Hand statue of the Dying Gaul in the entrance hall of Wentworth Woodhouse, wavenumber shift range 100–2800 cm^{-1}, laser excitation 785 nm, showing the evidence for it being of a cementitious plaster composition with Raman signals for anatase, gypsum, anhydrite and dolomite with no evidence for it being a fired ceramic

Stourhead, Syon Park and Wolterton Hall gave rather different results as described below:

Wentworth Woodhouse: The Right Hand statue of the Dying Gaul, whose Raman spectrum clearly shows bands that are assignable to gypsum, anhydrite, calcite and dolomite and that there is no supportive evidence of a fired ceramic. Likewise, the Left Hand statue of the Dying Gaul also exhibits band due to calcite and dolomite with bands of rutile from an unfired clay additive. This is of a different composition to the analogous statue described earlier and could originate from a type of cement to which a cold clay has been added.Clearly, both statues cannot be classified as Coade stone in origin and are probably of a later construction. The presence of a lead white signal in the spectrum of the Left Hand statue also indicates that a post-constructive treatment of a white pigmented wash has been applied to the figure at some stage.

Stourhead: the Borghese Vase analysis by Raman spectroscopy clearly indicates that it is of a fired clay composition with the additional presence of gypsum arising from a restoration procedure.

Wolterton Hall: a rather poor spectrum indicated nevertheless that it was of an unfired composition and exhibited Raman bands arising from gypsum.

Syon Park Gateway: the Raman spectrum was indicative of an unfired clay and quartz composite so could not be firmly assigned to Coade stone and the specimen probably originated from either the earlier artificial stone of George Davy or perhaps from a later cementitious restoration.

These molecular spectroscopic results reinforce the conclusions already made about the compositional variation in the specimens from different locations, several of which are assignable to a high-temperature fired ceramic such as Coade stone and others not.

References

C.C. Aggarwal, C.K. Reddy, *Data Clustering Algorithms and Applications* (CRC Press, Boca Raton, Florida, 2014), p. 89. *et seq*

A. Bezur, F. Casadio, The analysis of porcelains using hand held and portable XRF spectrometers, in *Handheld XRF for Art and Archaeology*, ed. by A.N. Shugar, J. Mass, (Leuven University Press, Leuven, 2012), pp. 249–311

J.M. Blashfield, *A Catalogue of Seven Hundred Articles Made in Patent Terra Cotta, and Red and Cane-Coloured Pottery* (Henry Johnson, Stamford, 1882)

B. Borg, M. Dunn, A. Ang, C. Villis, The application of state-of-the-art Technologies to support artwork conservation: a literature review. J. Cult. Herit. **44**, 239–259 (2020)

C.J. Brooke, Discovering groups in ecclesiastical data: the use of cluster analysis in the Southwell and Nottingham Church History Project, in *Church, Land and People, Essays Presented to John Beckett*, (The Thoroton Society, Nottingham, 2020), pp. 209–228

C.J. Brooke, H.G.M. Edwards, F. Košek, J. Jehlička, C. Stanford, M.J. Champion, The analysis of two historic fired artificial stonewares: Coade stone (18th–19th centuries) and a recent discovery from Oxnead Hall, Norfolk (16th–17th centuries). Archaeometry **65**(4), 691–938 (2022)

C.M. Cable, J.W. Smedley, Liquidus temperatures and melting characteristics of some early container glasses. Glass Technol. **28**, 94–98 (1987)

M. Charrad, N. Ghazzali, V. Boiteau, A. Niknafs, NbClust: an R package for determining the relevant number of clusters in a data set. J. Stat. Softw. **61**(6), 1–36 (2014)

S.A.H. Church, *English Porcelain: A Handbook to the China Made in England During the 18th Century as Ilustrated by Specimens Chiefly in the National Collection* (A South Kensington Museum Handbook, Chapman & Hall Ltd, London, 1885) and 1894

J. Cowan, *The George III Coade Stone Statue, Lincoln Castle,* Report prepared for the Skillington Workshop Ltd. and the Lincolnshire Conservation Laboratories (County Archives) (December 2008)

J.P.S. Davis, *Antique Garden Ornaments – 300 Years of Creativity* (Antique Collectors' Club, Suffolk, 1998)

S.H. Davy, Some experiments and observations on the colours used in paintings by the ancients. Philos. Trans. R. Soc. **105**, 97–124 (1815)

L.W. Dillwyn, *Notes and Workbooks of Recipes at the Swansea China Works*, 1815–1817, reproduced in H. Eccles & B. Rackham, *Analysed Specimens of English Porcelain in the Victoria & Albert Museum Collection* (Victoria & Albert Museum Publications, London, 1922), and in Edwards, *Nantgarw and Swansea Porcelains; An Analytical Perspective* (Springer, Dordrecht, 2018)

H. Eccles, B. Rackham, *Analysed Specimens of English Porcelain in the Victoria & Albert Museum Collection* (Victoria & Albert Museum Publications, London, 1922)

H.G.M. Edwards, *Nantgarw and Swansea Porcelain: A Scientific Reappraisal* (Springer, Dordrecht, 2017)

H.G.M. Edwards, *Nantgarw and Swansea Porcelains; An Analytical Perspective* (Springer, Dordrecht, 2018)

H.G.M. Edwards, *Porcelain to Silica Bricks: The Extreme Ceramics of William Weston Young, 1776–1847* (Springer, Dordrecht, 2019)

H.G.M. Edwards, *Porcelain Analysis and Its Role in the Forensic Attribution of Ceramic Specimens* (Springer, Dordrecht, 2022)

H.G.M. Edwards, P. Vandenabeele, P. Colomban, *Raman Spectroscopy in Cultural Heritage Preservation* (Springer-Nature, Cham, 2023)

R.A. Ferchault de Réaumur, *Idee generale des diferente matieres dont on peu faire la porcelaine et quelles sont les veritables matieres de celle de Chine* (Memoires Academie des Sciences, de Paris, 1727)

R.A. Ferchault de Réaumur, *Second memoire sur la porcelaine ou suite des principes qui douvent conduire dans la composition des porcelaines de differente genres et qui etablissent les caracteres des matieres fondante qu'on ne peut choisir pour tenir lieu de celle qu'on emploie a la Chine* (Memoires Academie des Sciences de Paris, 1729)

R.A. Ferchault de Réaumur, *Memoire sur l'art de faire une nouvelle espace de porcelaine par des moyens extremement simples et fáciles ou de transformer le verre en porcelaine* (Memoires Academie des Sciences, de Paris, 1739)

W.H. Finch, J.E. Bolin, *Multilevel Modeling Using R*, 3rd edn. (CRC Press, Boca Raton, 2024), p. 289

C. Fowler, Some remarks on Terracotta and artificial stone as connected with architecture. Civil Eng. Archit. J. **13**, 215–216 (1850)

I. Freestone, Forgotten but not lost: the secret of Coade stone. Proc. Geol. Assoc. 135–138 (1991)

I. Freestone, M. Bimson, M.S. Tite, The constitution of Coade stone, in *Ancient Technology, Chapter 10: Modern Science, Ceramics and Civilisation*, ed. by W.D. Kingery, vol. I, (American Ceramics Society, Columbus, 1985), pp. 293–304

F.H. Garner, M. Archer, *English Delftware (Monographs on Pottery and Porcelain)* (Faber & Faber, London, 1972), p. 58

P. Giordani, M. Brigida Ferraro, F. Martella, *An Introduction to Clustering with R, Part II Standard Clustering* (Springer Nature, Singapore, 2020)

S.B. Hamilton, Coade stone. Archaeol. Rev. **116**, 295–301 (1954)

W.K. Härdle, L. Simar, *Applied Multivariate Statistical Analysis*, 5th edn. (Springer Nature, Cham, 2019), pp. 299–336

L.L. Harlow, *The Essence of Multivariate Thinking* (Routledge, New York, 2023), pp. 154–155

C. Hitching, *Rock Landscapes: The Pulham Legacy* (Antique Collectors' Club, Suffolk, 2012)

D.A. Holdridge, Ball clays and their properties. Br. Ceram. Soc. Trans. **55**, 369–440 (1956)

M. Hornig-Sutter, Meissen Porzellane des18 Jahrhunderts im Visier von Kunstgeschichte und Naturwissenschaft. Teil I. Weltkunst **54**(23), 3654–3658 (1984)

I.T. Jolliffe, *Principal Component Analysis*, 2nd edn. (Springer, New York, 2002)

L.E. Karran, Ph.D Thesis (University of Lincoln, 2023a)

L. Karran, Personal communication (2023b)

L.E. Karran, B.J. Colston, Coade, Blashfield or Doulton? The in situ identification of ceramic garden statuary and ornaments from three eighteenth and nineteenth manufacturers. J. Cult. Herit. **18**, 290–298 (2016a)

L.E. Karran, B.J. Colston, Scientific investigation of the Shugborough Druid. Gard. Hist. **44**, 150–152 (2016b)

A. Kelly, *Mrs Coade's Stone* (Self-Publishing Association, Worcestershire, 1990)

A. Kelly, Coade stone: its character and conservation, in *Architectural Ceramics: Their History, Manufacture and Conservation, A Joint Symposium of English Heritage and the United Kingdom Institute for Conservation, 22nd–25th September,1994*, ed. by J.-M. Teutonico, (James & James Ltd, London, 1996), pp. 32–38

B. Lei, G. Xu, M. Feng, Y. Zou, F. van der Heijden, D. de Ridder, D.M.J. Tax, *Classification, Parameter Estimation and State Estimation* (John Wiley, Chichester, 2017), pp. 323–329

P.D. McNicholas, *Mixture Model-Based Classification*, Chapter 2 (Taylor and Francis, Boca Raton, 2016)

E. Morton Nance, *The Pottery and Porcelain of Swansea and Nantgarw* (Batsford & Sons, London, 1942)

M. Ponting, Investigation of Coade stone fragments from West Dean College, Sussex, in *Centre for Archaeology Report*, (English Heritage, Portsmouth, 2003), pp. 1–5. ISSN: 1473-9224

W.R.H. Ramsay, H.G.M. Edwards, E. Manners, A. Howkins, Geochemical Investigation of a Ceramic Snuff Box – Body, Glaze, and Enamels: 'A' – Mark English Porcelain Attribution Confirmed, in *Ceramics in America* (in press, 2023)

L. Skerrit, *Condition Report and Assessment of the Immediate and Long-Term Needs of the Coade Stone Statue of George III.* Unpublished Report (Lincolnshire County Council, 2006)

C. Stanford, personal communication to authors (2024)

A. Stiegelschmitt, G. Tomandl, Untersuchung der Materialzusammensetzung alter Porzellane Im Hinblick auf ihre kunsthistorische Zuordnung. Sprechsaal **118**(10), 974–978 (1985)

J. Thomas, The lesser known building materials of West Dorset. Proc. Geol. Assoc. **101**, 289–301 (1990)

M.S. Tite, M. Bimson, I.C. Freestone, A technological study of Fulham stoneware. In: Proceedings of the 24th Archaeology Symposium (Washington DC, USA, 1986) pp. 95–104

N. Valpy, A. Kelly, Advertisements for artificial stone", Trans. Engl. Ceram. Circ. 206–226 (1986)

H. van Lemmen, *Architectural Ceramics* (Shire Books, Princes Risborough, 2002)

H. van Lemmen, *Coade Stone* (Shire Books, Princes Risborough, 2006)

P. Vandenabeele, J. van Pevenage, Raman spectroscopy and the study of ceramic manufacture: possibilities, results and challenges, in *The Oxford Handbook of Archaeological Ceramics Analysis*, ed. by A.M. Hunt, (Oxford University Press, Oxford, 2017), pp. 531–543

R. Vidal, Y. Ma, S.S. Sastry, *Generalized Principal Component Analysis* (Springer, New York, 2016)

S.T. Wierzchoń, M.A. Kłopotek, *Modern Algorithms of Cluster Analysis* (Springer, Cham, 2018), p. 30

W. Yao, S. Xiang, *Mixture Models Parametric, Semiparametric, and New Directions* (CRC Press, Boca Raton, 2024)

Chapter 7
Case Studies: Coade Stone

Abstract Twenty-five case studies (see keywords below) of artificial stone locations with artefacts are each considered historically and reviewed on the basis of their potentially being genuine Coade stone or not: several are damaged artefacts and in other cases the original artefact is now missing. For several of these specimens have been obtained for analysis herein and these results are generally indicative of the genuineness of the article concerned. In some cases the situation is still uncertain because of the later restoration that has been effected which has compromised the conclusion and in other cases the result is in abeyance. Some case studies would certainly repay analytical sampling to determine the outcome.

Keywords Case studies · Coade stone · Belton House · Lion of Huddersfield · Austin & Seeley eagle · South Bank lion · Parham House · Dunston Pillar · King George III Lincoln Castle · St George's Chapel Windsor · Anglesey Abbey · St Pancras Church · Farnley Hall · Kew Gardens · Stourhead · Castle Howard · Wentworth Woodhouse · Dying Gaul · Harlaxton Manor · Witley Court · Malvern Hall · Lord Schomberg's House · King George III Weymouth · Mrs Gandes · Nelson's Columns · Lord Hill Shrewsbury · Weston Park · Wolterton Hall · Syon Park Gateway

The number of manufactories that were engaged with the production of artificial stone monuments, statuary and architectural detail after Eleanor Coade's business closed down in 1837 has been discussed earlier and this is surely a result of the great success that she made in promoting the idea and effectively realising its full potential in Georgian society through her resilient and degradation-defeating fired ceramic artefacts. A proliferation of Coade stone replica companies, others who made ceramic variants and others based especially upon the new cementitious cold-setting mixtures that did not require many days' firing in kilns at elevated temperatures then appeared around the mid-nineteenth century, such as Blashfield and Blanchard for the former category and Austin & Seeley for the second category. The current situation regarding attribution is subject to some confusion in identifying artefacts by the continued use of moulds that had been either directly acquired from the winding-up sales of the Coade business in 1843 or through their alternative

H. G. M. Edwards, C. J. Brooke, *Coade Stone*, Cultural Heritage Science,
https://doi.org/10.1007/978-3-031-71432-0_7

manufacture using new secondary moulds that were made from existing and genuine Coade stone garden statuary that had been originally provided to the owners of the great houses and estates. This means that it is no longer correct to assume that a statue or an artefact that looks as if it is made of Coade stone, and possibly even of an identical or of a very similar design, is an original product of Eleanor Coade's factory in Lambeth between 1769 and 1837.

During their research carried out for this book, the authors have come across several instances where an artefact has been attributed to Coade stone on the sole basis of a casual visual historic observation and it is realised that these should now be open to verification and possibly then be the subject of an alternative interpretation and assignment. In some cases, the documentation required for a proper provenancing and dated original invoices still exist in country house archives which confirm the order of a particular statue or ornament provided from Eleanor Coade, John Sealy or William Croggon, often citing the number of the artefact as presented in the Coade and Sealy Gallery Catalogue of 1784, and these indeed give valuable clues to the existence of the artefacts concerned somewhere in the property or estate. Of course, as estates fell upon hard times their assets were often sold off at auction and were assimilated by neighbouring estates and by other clients, thereby suffering a potential loss of provenance which annuls the pursuance of the forensic audit trail associated with that original artefact historically. Alison Kelly, in her comprehensive text on Coade stone artefacts (Kelly, *Mrs Coade's Stone*, 1990), frequently comments that a particular item that had been observed in a specific location historically, and for which there may have been a documentary record in support of its origin and date of commission originally, is no longer visible at that location: the phraseology used by Alison Kelly in such cases is either "not identified" or "not traced". These observations are recorded under the headings of Questionable Authenticity (230 items) or Destroyed /Damaged (248 items) in the numerical analysis summary data given here in Table 1.2 and represent some 18% and 20%, respectively, of all the Coade stone pieces surveyed in Kelly's text. This translates currently to over one third of known and listed Coade stone pieces that fall into these two categories, which is not an insignificant number for our survey. In other cases encountered by the present authors, the curators of some estates do not know if they still have Coade stone artefacts in their current possession, even when the house archives as cited by Alison Kelly originally indicated that they should have. Care must also be taken with the strict interpretation of William Croggon's day books as the quotation that he provided for a specific item's manufacture for a client does not necessarily mean that this was actually made to a commission and subsequently delivered on site to the client. In several cases, it is clear that the commission was never undertaken and that the items were never delivered: in one case, at least, the potential client has been mis-identified and was invoiced for Coade stone artefacts that were not received and it must be presumed that these were delivered to another client incorrectly elsewhere, prompting an angry exchange of correspondence between William Croggon and the wrong client!

The following selected case studies cover several aspects of Coade stone artefacts, real and imaginary, that have at some time all been classified as having been manufactured from genuine Coade stone and all are characterised wherever possible by the application of a modern holistic forensic investigative approach which in its widest mantra covers connoisseurship opinion, historical documentation and scientific data where applicable, to demonstrate a more credible attribution; an established procedure which has had a demonstrative success in its application to the analogous ceramic field of porcelains and their attributed factories of manufacture which on the basis of connoisseurship opinion alone is often fraught with errors and difficulties.

7.1 Belton House (Belton), Lyme Regis, Dorset

A Georgian manor house, Belton House (or more simply Belton), was bequeathed to Eleanor Coade by her uncle Samuel Coade in 1787, although she had been leasing it from him since 1784, and she immediately embellished it and adopted it as a template for her own architectural additions such as keystones, architraves, voussoirs and ornamental urns (Fig. 7.1). It is currently in the ownership of the Landmark

Fig. 7.1 Belton House (Belton), Lyme Regis, Dorset, the country home of Eleanor Coade since her leasing it from her uncle Samuel Coade in 1784, when it was known as Bunter's Castle. She adorned the house with applied architectural features in her Coade stone, such as architraves, keystones, voussoirs and ornamental urns. The gatepost piers have applied paterae in Coade stone and her crest, a coot (which is a pun on her name as pronounced in the local dialect), is seen on the right hand pier

Trust, since 2007. Two samples of Coade stone from the gatepost paterae were analysed earlier by the current authors and the results which appeared in an earlier paper (Brooke et al. 2022) are reported alongside other exemplars in this book (Chap. 6). Eleanor Coade's occupancy of Belton House has given rise to a mistaken assumption by some historians as the reason for her use of Dorset clays in the formulation recipe of Coade stone; this was based on the incorrect premiss that she was born in Lyme Regis, Dorset, when in fact she was born in Exeter, Devon, and did not reside in Dorset until she inherited the house in 1787—some 18 years after her foundation of Coade & Co., and when the manufactory was in full scale production.

Belton House was established as a an elegant two-storey seaside pavilion by Simon Bunter around 1772 and was known thereafter as Bunter's Castle. It was bought by Samuel Coade, a wealthy clothier and fuller in 1783, who transferred the lease to his niece Eleanor Coade in 1784. Eleanor then refurbished and embellished the property as indicated above and retained it until she died in 1821. From 1824 until 1881 there were a succession of tenants and a Mrs. Fewtrell notably renamed the house Belton House, or more simply Belton, in the mid-nineteenth century. Belton became the residence of a Dr. Richard Bangay between 1881 and 1896 and thereafter by a Dr. Harold Cooper and his descendants until 1968, when it was purchased by the author and novellist, John Fowles. It may be inferred therefore that Belton House during Eleanor Coade's residence was still known officially as Bunter's Castle.

7.2 The Lion of Huddersfield

Whilst researching local examples of Coade stone statuary in the north of England, of which there seem to be relatively very few listed in Alison Kelly's book, a chance encounter was made with the "Lion of Huddersfield". Although on the surface this provided another example of a potentially long-lost Coade stone artefact that has hitherto escaped discovery but which has now been recognised, the detailed research actually revealed a more complex story which creates some doubt about its official attribution as a Coade stone statue. The civic walks conducted by local historians in Huddersfield in West Yorkshire mention the fine Victorian urban architecture constructed of local quarried sandstone and in particular the presence of a large lion statue situated atop the Royal Arcade, also known as the Lion Chambers, on John William Street and in St George's Square, which were constructed by Samuel Oldfield according to the design of the architect James Pigott Pritchett under commission to Sir James Ramsden and his architectural consultant, William Tite, in 1852. St Georges's Square has been described as an outstandingly fine example of a provincial Victorian square. The entry for its Historic England listing of 29th September 1978 (entry no. 1134167) gives the following statement:

JOHN WILLIAM STREET (East Side). Lion Buildings. 1853. Architect J P Pritchett. Ashlar sandstone. Hipped slate roof. Three storeys. Composed with rounded and pavilions, central pavilion breaking forward and a central frontispiece. This has coupled columns on each floor (attached Tuscan on ground floor, detached composite above), taking pediment and parapet with entablature, paired pilasters, and ***giant free-standing coade stone statue of lion by Seeley****. The corners of each pavilion are marked on first floor by paired attached composite columns (outer ones square), as is the central pavilion alone on the second floor also. The first floor also has paired attached composite columns in middle of wall between each pavilion, and two single columns spaced regularly round curve of end pavilions. On St Peter's Street elevation there are three pairs of Tuscan pilasters, one at corner with Wood Street and two others flanking central group of windows. Ground floor has modern shopfronts separated by rusticated and vermiculated pilasters, arched central doorway with moulded voussoirs and imposts, and keystone with dragon and crown inscribed. Dentilled cornice. On Northumberland Street elevation a mezzanine is inserted below ground floor cornice, consisting of sashes framed by Tuscan pilasters according to the following rhythm: major pilaster-3-major pilaster-1-minor pilaster-1-minor pilaster-1-major pilaster-3-major pilaster. First floor has full entablature which breaks forward and is dentilled over each pair of columns, and the windows of the frontispiece are provided with a stone balcony on paired scrolled consoles. Windows are round-arched sashes with moulded voussoirs, and are separated by Ionic colonnettes: they are grouped in the following rhythm, Northumberland Street elevation: 1-1-1-2 (crowned by pediment) -1-1-1. End pavilion: 2-3-2. St George's Square elevation: 4-4. Central pavilion: 4-3 (central one slightly wider and higher) -4. St George's Square elevation (cont): 4-4. End pavilion: 2-3-2. St Peter's elevation: 2-1-2-1-2. Second floor has modillioned entablature, save at end pavilions where modillions become consoles. Fenestration is same as first floor, save on St George's Square elevation where grouping is as follows: 2-1-2-2-1-2-2-1-2. Some semblance of order is given to this otherwise chaotic design, but only above the eaves cornice. Here the end pavilions have a solid parapet with cornice (inscribed "Lion Arcade" in relief), piers with ornamental caps and huge flanking ornamental scrolls, while central pavilion has balustrade with moulded rail, pairs of square Tuscan columns with ornamental caps at ends, and square Ionic balusters. Over the frontispiece comes the parapet with "1853" in relief and the lion on top.*

The statement on line 5 is most revealing in that it says "... *and giant free-standing coade stone statue of lion by Seeley*", which has been highlighted above. Also, the local history archives state that the original lion statue was replaced in 1977 by a fibreglass version because of its "*severe deterioration*" ... both of these statements need further analysis because at face value they imply that a Coade stone statue has in fact not survived in its environment and exhibited very serious deterioration, so providing perhaps a rare and visibly documented exemplar of that situation recorded historically! Local legend has it that the lion gets down from his plinth and walks around the square when the station clock strikes 12: however, the station clock is a non-striking one ... so this magical occurrence has never been witnessed to date!! Another local historian has maintained in his talks that the lion was made of concrete strengthened with iron struts, and this is also now worthy of consideration as it implies that it was possibly not even made of Coade stone but perhaps later was fashioned from a cold-cured cement? A photograph of the Lion of Huddersfield sitting on the Royal Arcade buildings is shown in Fig. 7.2: this is estimated from the historical archive to date from the early 1960s. A slightly later picture (Fig. 7.3), dating it is estimated from perhaps the early 1970s, shows that the lion statue has indeed deteriorated, and that it has shed some of its outer surface (spallation or delamination) and is now potentially in a dangerous condition. A little later still, this deteriorated stone lion statue was removed and was replaced with a fibreglass version by Kirklees Council in 1977.

The following points about this questionable Coade stone artefact can now be made:

- The Historic England statement reproduced above states that the lion was ordered from someone called Seeley—and the clear implication that derived from that by early historians was that this was John Sealy (authors: however, note the spelling

Fig. 7.2 the Lion of Huddersfield atop the pediment of the Royal Arcade Buildings on John William Street, St George's Square, Huddersfield with the date 1853 beneath; photograph taken approximately around 1968

Fig. 7.3 A close-up of the Lion of Huddersfield taken in the 1970s showing deterioration and exfoliation of the surface: it was removed and replaced with a fibreglass version a little later

change here!), who was the partner in Eleanor Coade's artificial stone business from 1771 until he died in 1813. Up until this time the firm had labelled itself as Coade & Sealy, Lambeth, to reflect his active business partnership with Eleanor Coade.

- As John Sealy had died in 1813, some 40 years before the Huddersfield project started, it would have been impossible for the architects of the Royal Arcade in Huddersfield to have commissioned a lion statue from him personally. Also, the successors to Eleanor Coade's business were firstly William Croggon who carried on with the business until 1833, then his son, Thomas Croggon, who took over in 1835 and finally the business was acquired by Routledge & Co. after 1838, who eventually finally sold it off at auction through Rushworth, Jarvis & Co of Savile Row in July, 1843. At this stage the moulds were purchased by aspiring artificial stone manufacturers such as Mark Blanchard and John Blashfield, who carried on their business of manufacturing artificial stone in the style perfected by Eleanor Coade until the 1880s.
- Hence, it is perhaps illusory and certainly potentially very misleading to connect John Sealy (*sic.*) and the Huddersfield Lion with Coade stone, which had finished its artefact production in Lambeth prior to 1840. So, we must now try to establish the identity of this person Seeley, who is mentioned in the Historic England statement given above as being the supplier of the lion statue.

- A search through the literature reveals that the firm called Austin & Seeley certainly existed in 1840, when a John Seeley had joined the firm founded by Felix Austin some years before, directly from the Royal Academy, to make a cold-cured cementitious artificial stone: they marketed their product as "everlasting stone" and could offer it more cheaply than Coade stone as it was cold-cured at ambient temperatures and was not therefore a high-temperature fired ceramic. By the early 1850s they were well established in this business and producing garden statuary and ornaments prolifically in the neo-classical Roman, Greek and ancient Egyptian styles reminiscent of those manufactured earlier in Coade stone.

7.2.1 Austin & Seeley

In 1800 a Dutchman, van Spangen, in partnership with a Mr. Powell set up an artificial stone manufactory at Bow in London; the firm was not successful and closed in 1828, when Felix Austin then bought the moulds and set up a manufactory in New Road, London, now the Euston Road. He took into partnership John Seeley, a bright new architect, statuary mason and sculptor fresh from the Royal Academy to manufacture artificial stone based upon the new discovery of Portland cement by Joseph Aspdin and his son of Leeds that was announced earlier in 1828. Austin & Seeley's formulation comprised Portland cement, broken stone pieces, powdered marble and a coarse sand which did not require a vigorous thermal treatment to cure—in effect, this was the first recorded commercial application of a cementitious artificial stone (*The Builder* 1868) adopted for garden statuary. In 1841, Austin & Seeley prepared their extensive Catalogue, known by its abbreviated title of "*Pleasure Grounds, Etc.*", whose full title was "*Collection of Ornaments at Austin and Seeley's Artificial Stone Works for Gardens, Parks, and Pleasure Grounds, Etc.*", which displayed their repertoire of designs, many of which were based upon the earlier exemplars offered by Eleanor Coade (Scott, *Artificial Stone: A Successful Substitute for Natural Stone*, 2016). The interesting preface to the contents of this Austin & Seeley Catalogue is stated below:

> Austin's Artificial Stone is of a light tint, requires no painting or colouring, will not sustain injury from the severest winter, and, being impervious to wet, is particularly applicable to all kinds of water works. Its superiority is now so thoroughly established, that the most eminent architects and scientific gentlemen have expressed, in their highest terms, their approbation of its durability, and close resemblance to the real stone.

The Catalogue was published annually thereafter until the works closed in 1872. Austin & Seeley stressed the superiority of their product and accentuated its relative cheapness when compared with the technically more difficult to produce high temperature fired ceramic Coade stone. Austin & Seeley certainly made their mark quickly and are cited in Loudon's encyclopaedic survey of 1842 (Claudius Loudon, *Encyclopaedia of Cottage, Farm and Villa Architecture and Furniture: Containing Numerous Designs for Dwellings from the Cottage to the Villa including Farms and*

Other Agricultural Buildings and the Appropriate Office and Garden Furniture, 1842).

From the examples of Austin & Seeley's work on garden statuary that have survived it is clear that their claim for longevity of their product is exaggerated: the major problem in this context is associated with the porosity of the cementitious materials they used in construction to moisture both from within and without the structure and the degradative effect this has had upon the iron struts that were used for the internal reinforcement of the artefact: the corrosion of the inner iron struts is believed to be the major source of deterioration of Austin & Seeley's statuary. An example of an Austin & Seely eagle garden statue is shown in Fig. 7.4, showing damage to its right leg caused by corrosion of the inner iron strut support: Fig. 7.5 shows the same eagle after modern restoration and conservation. Certainly, many of the models and shapes of Austin & Seeley's statuary are copies of Eleanor Coade's artefacts; so precisely do these match superficially that it was at first suggested that Austin had purchased the Coade stone moulds but it is now thought that John Seeley, an accomplished sculptor, simply made moulds from existing Coade stone artefacts,

Fig. 7.4 An Austin and Seeley garden statue of an eagle made of cementitious stone, *ca.* 1850, with visible damage to its right leg caused from corrosion of the internal iron support struts

Fig. 7.5 The Austin and Seely eagle shown in Fig. 7.4, cleaned restored and sited in the garden

such as the Pineapples at Ham House, Sabrina at Croome Park and the River God, Father Thames. The Huddersfield Lion, weighing some 7 tons and being 11 feet long, would have been no exception to this venture. It is also recorded that Felix Austin persuaded landowners with Coade statuary in their estates to allow him access for copying purposes and several did so.

Enquiries made of the Kirklees Council by the authors, whose Conservation Department administer the historic buildings of Huddersfield, revealed that there was no record of what had happened to the artificial stone lion that had been removed from its Royal Arcade plinth in 1977. This is unfortunate because chemical analysis of a damaged specimen from this statue should be able to reveal that, firstly, it is perhaps not made of Coade stone as alleged, and, secondly, that it analyses as a cold-cured cementitious stone with an abnormally and appropriately high calcium content, *viz.*, some 30% or more CaO compared with its Coade stone fired ceramic analogue at about 1% or less. Also, inspection of the degraded fragments of the Huddersfield Lion should reveal mineralogical domains that are rich in iron (III) oxide and iron salts from corrosion of the internal iron struts, and possibly some molecular signatures that could be ascribed to mineral alteration associated with acid rain pollution and also atmospheric sulfur or nitrogen oxide atmospheric contamination from Victorian industrial times. A further approach made to the West Yorkshire Archaeology Service and to the Historic Environment Records Office was more successful and a summary of the information generated is given in Appendix IV, including some interesting close-up photographs taken of the Lion in 1976 for assessment of its stability before removal of the sculpture from its plinth and pediment was undertaken (Figs. AIV.1, AIV.2, AIV.3, AIV.4, AIV.5, AIV.6, and AIV.7).

7.3 The South Bank Lion

No discussion of Coade stone artefacts would be complete without a mention of the South Bank Lion, certainly one of the largest Coade stone artefacts ever made, at 12 feet in length and probably the heaviest at 14 tons, of the surviving Coade stone relics of the era: Fig. 7.6 shows the South Bank Lion in its current location on a plinth abutting Westminster Bridge, London. We have already commented on this artefact in some detail in earlier chapters and how it has survived several relocations from its original setting atop the main entrance to the Lion Brewery in Lambeth in 1837. The Lion Brewery was founded in 1836 on land owned by the Archbishop of Canterbury by James Goding, who commissioned three lions in Coade stone for their buildings in 1837: the largest sat on top of the main building overlooking the River Thames and Belvedere Road adjacent to the Hungerford Bridge and situated

Fig. 7.6 The South Bank Lion—originally made for the James Goding Lion Brewery in Lambeth in 1837, bearing the script mark *WFW Sculptr COADES 24th May 1837*, which was discovered when it was relocated from its original site to the concourse of Waterloo station in 1949

Fig. 7.7 Early photograph of River Thames frontage of the Lion Brewery with the Coade stone Lion in position on the roof

between that and the Charing Cross Bridge: Fig. 7.7 shows the Lion Brewery in an early photograph with the lion atop its main facade and facing the River Thames and also in a contemporary Victorian print of a waterfront scene, showing the shot tower between the Lion Brewery and Waterloo Bridge (Fig. 7.8). It became known as the Lion Brewery Co. in 1866. The Lion Brewery was a major exporter of strong beer to the British Empire, carried by the Honourable East India Company ships, and was called India Pale Ale (Fig. 7.9) with its characteristic lion logo. In 1924 the company was taken over by Hoare & Co. of East Smithfield, Wapping. The Brewery suffered a serious fire in 1931 and closed thereafter, the site then becoming derelict until its demolition in 1949 to provide space for the Festival of Britain and the Royal Festival Hall (Roberts & Godfrey, "Lambeth Waterworks and the Lion Brewery", in *Survey of London: Lambeth: South Bank and Vauxhall*, 1951). The large lion was relocated to the Waterloo station concourse in 1949 and painted red to match the logo of the then British Rail (Fig. 7.10), before being moved finally to its current site on

Fig. 7.8 Print of a view of the River Thames waterfront between Hungerford Bridge and Waterloo Bridge, *ca.* 1850, showing the Lion Brewery main façade topped by the Coade Stone lion shown in Fig. 7.6. The buildings to the right and rear of the Lion Brewery are the Lambeth waterworks, and the tower to the left is the Shot Tower

Fig. 7.9 The lion logo of the Lion Brewery adopted for its India Pale Ale adopted from 1866

Fig. 7.10 The Lion Brewery statue from Fig. 7.9 in its new location on the concourse of Waterloo Station painted red to match the British Rail logo behind it

Westminster Bridge to permit an extension to the station concourse to be accomplished. It is alleged that the large lion was assembled in parts of Coade stone and cramped together using iron hoops, which was certainly a departure from the normal Coade manufacturing practice for smaller statues of firing each artefact as a complete entity. It should be noted, however, that the large statue of King George III from Dunston Pillar (originally some 5 metres tall) and now in fragments in Lincoln Castle exhibits bronze and ferrous dowels secured with lead casts in several broken pieces, indicating that a metallic reinforcement of parts of the moulded paste figure had been effected: a similar observation relating to the metallic strengthening of some 12 pieces was made during the assembly of the figure of Lord Clive (height 5.2 metres) on his column in Shrewsbury. A second smaller Coade stone lion, which was based upon an entablature over another entrance to the Brewery (Fig. 7.11), now graces the Rowland Hill stand entrance to the English Rugby Union Football headquarters at Twickenham Stadium (Fig. 7.12). The third lion from the Lion Brewery site had disappeared before the demolition of the site had started in 1949,

Fig. 7.11 The second lion of the Lion Brewery guarding the rear entrance on Belvedere Road

Fig. 7.12 The second lion originally at the Lion Brewery (Fig. 7.11) in its relocated position outside the Rowland Hill gate at the headquarters of the English Rugby Union Football Club, Twickenham. It was gilded in 1991 to celebrate the World Cup being hosted at Twickenham stadium

Fig. 7.13 The missing third lion from the Lion Brewery standing above the entrance to the brewery building at the corner of Belvedere Road and Sutton Walk. Photograph probably dating from the late 1920s or perhaps the early 1930s, as the lion had disappeared after the fire in 1931 which caused the site to become derelict and before it was cleared in 1949 for The Festival of Britain

but an old photograph (Fig. 7.13) depicts this lion above the entrance to the second building of the Lion Brewery at the corner of Belvedere Road and Sutton Walk. The Lion Brewery has since been resuscitated in Singapore in 2018 and now markets Straits Pale Ale but still retains its lion logo on the label. It is possible that one of these larger Coade stone lions formed the template for the moulding of the alleged Austin & Seeley version of the Lion of Huddersfield discussed earlier.

An old plan of the Lion Brewery buildings situated on its River Thames frontage between Hungerford Bridge and Waterloo Bridge is shown in Fig. 7.14, with the designated positions of the three Coade stone lions marked as 1, 2 and 3.

Finally, King George VI decreed that the South Bank Lion should be recognised as an emblem of the stoicism and courage shown by the British people during the darkest days of World War II and he commissioned a plaque to this effect which would be placed on its plinth on Westminster Bridge upon its relocation there finally in 1966 (Fig. 7.15).

7.4 Coade Stone Thrones

The neo-classical theme of much of Eleanor Coade's garden statuary was derived from several volumes of sketches that were made by earlier chroniclers and it is recorded that the Coade Gallery had access to the works of Bottari (*Il Museo*

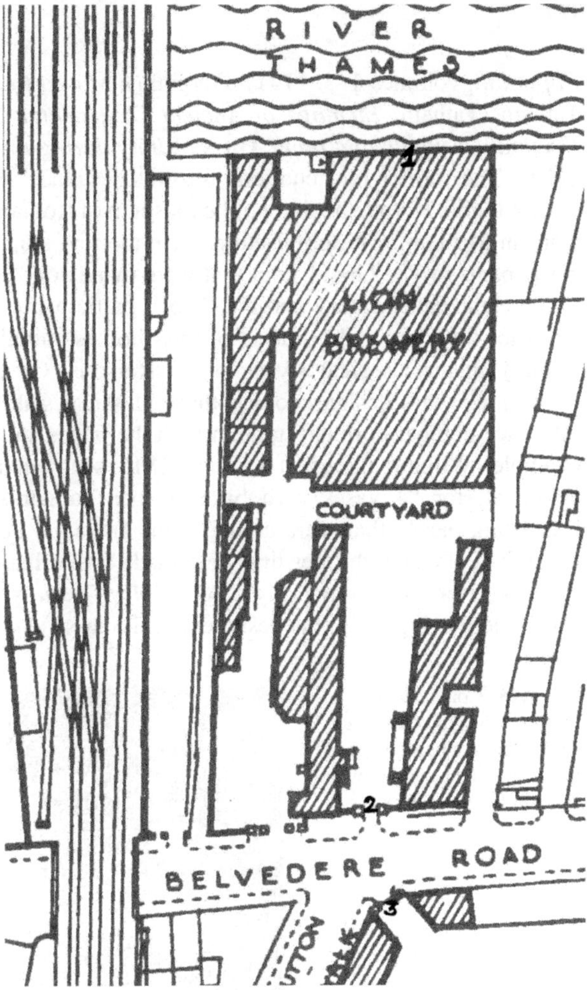

Fig. 7.14 Map taken from Roberts and Godfrey's *Survey of London: Lambeth, South Bank and Vauxhall* and modified to show the placements of the lions 1, 2 and 3 relative to each other at the Lion Brewery site

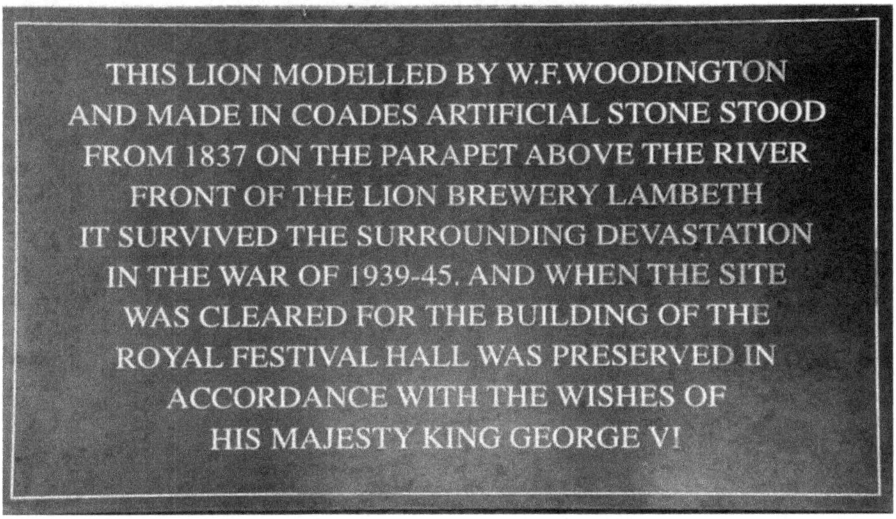

Fig. 7.15 Plaque affixed to the plinth of the South Bank Lion on Westminster Bridge

Capitolini, volumes 1–3, 1741), Foggini (*Il Museo Capitolini*, volume 4, 1750) and Tatham (Tatham, *Etchings of Ancient Ornamental Architecture drawn from the Originals in Rome and other Parts of Italy during the Years 1794,1795 and 1797,* 1803). The former two citations illustrated statues in the Capitoline Museums at Campidoglio, Rome, the first public museum of Roman art, and the latter those in the Vatican Museum. These museums were the first stopping point for the aristocratic participants in The Grand Tour that were mandatory for emerging Georgian society in the eighteenth century and the British participants returned enthused with novel ideas for the interior decoration of their homes and also for their new look garden statuary in the classical styles of the Romans, Greeks and Egyptians. Many of Eleanor Coade's designs copied these classical statues faithfully and even had a reference number correlating them with the plates and sketches in these books. For example, William Croggon in June 1818 sent out an invoice for £38-2s-0d to Whitehall for "4 vases with birds as per Tatham's book". Some of the most interesting pieces that were copied from Tatham's book of Vatican etchings were the pair of classical thrones that were made for Parham House, Sussex. A picture of these is given in Alison Kelly's book (*Mrs Coade's Stone*, 1990, page 94) and is shown here in Fig. 7.16: the designer of this classical stone throne was Thomas Hope

Fig. 7.16 Coade stone throne with sphinx arms, *ca.* 1800: one of a pair designed by Thomas Hope and now in Parham House gardens

(1769–1830), who was very influential upon Regency style and furnishings, especially with the inclusion of Egyptian motifs and symbols, as he believed in the importance of ancient Egyptian themes upon Western culture (Watkin, *Thomas Hope and the Neo-Classical Idea*, 1968; Watkin & Hewat-Jaboor, *Thomas Hope, Regency Designer and Patron*, 2008). Thomas Hope himself made the Grand Tour between 1787 and 1795 and included many of his ideas in neo-classical designs in his own publication, *Household Furniture and Interior Decoration* (Hope 1807), which proved to be so influential in the style of Regency house furnishings for the following decade: the designs of these thrones are illustrated in Plate 19 of his book, specifically figs. 6 and 7 given there. Hence, the classical thrones made by Eleanor Coade and which were then located at Parham House, as identified by Alison Kelly, are based on Thomas Hope's designs: there was some conjecture as to whether these are actually made of Coade stone or natural stone and this is recorded with some doubt in Alison Kelly's book (*Mrs Coade's Stone*, 1990, p. 94). Parham House was built in 1577 and bought in 1601 by Sir Thomas Bishopp, a naval colleague of Sir Francis Drake, whose descendants then lived there for over 300 years. The pleasure gardens at Parham contain many sculptures, including a River God, illustrated in Coade's Catalogue as item No. 1, which may be Coade stone or perhaps even a later replicate commissioned from her competitors. The thrones that are the subject of this case study were also shown located in the gardens at Parham in Kelly's book.

7.4.1 Thomas Hope and the Regency Style

Thomas Hope was born in Amsterdam in 1769 into a wealthy family of Dutch bankers, Hope & Co., and he embarked upon an extended Grand Tour lasting 8 years from 1787 until 1795, where he visited the classical sites in Italy, Greece, Turkey and Egypt, which inspired his admiration for classical statuary and was translated into the acquisition of a noteworthy collection of antiquities on a grand scale. On his return to the Netherlands in 1795 he found that Napoleon had invaded his country in January of that year, so Thomas exiled himself to London and was resident first at 2, Hanover Square, which he filled with antiques, sculpture and precious works of art from his travels. He then bought Clerk House in Duchess St, Portland Place, near the fashionable Cavendish Square, which had been built in a Parisian style (known as a "*hotel peculier*") by Robert Adam between 1768 and 1771 for General Robert Clerk and his wife, the Dowager Countess of Warwick. Clerk House was damaged by fire in 1771 and was rebuilt by Robert Adam, who completed his rebuilding task assignment in 1775. Whilst in Rome on his Grand Tour, Thomas Hope met with the architect and designer Charles Tatham, who was also in Rome on the Grand Tour between 1794 and 1796, and they formed a lasting friendship. It was Tatham who was engaged by Hope to re-design the interiors of the rooms of Clerk House in 1799 in the neo-classical style that was to become the signature of Hope: the interiors created by Hope in his Duchess St home to display his works of art were most influential with his Regency society circle and the discovery of drawings by Robert

Adam and Charles Tatham that were made in 1799 (Adam & Tatham, *Apollo*, 2004) bear witness to their sumptuousness of design and classical influence. Hope opened his home in Duchess St in 1802 to an illustrious gathering of artists, Royal Academicians, scholars and Royalty, which included Sir John Soane and the Prince of Wales. In his Egyptian Room, for example, he mixed genuine Egyptian antiquities with those he had created himself neo-classically in the Egyptian style. The Picture Gallery drawing is most relevant here since it shows clearly four Egyptian style thrones (Fig. 7.17) which were later copied by Eleanor Coade and manufactured in Coade stone for location in a generous space inside the house designed by Tatham that was 96 feet long and 24 feet wide.

Thomas Hope married Louisa van de la Poer Beresford in 1806 and thereby inherited a large country home at Deepdene, near Dorking in Surrey, a Georgian mansion that he refurbished and remodelled in an Italianate "loggia" style. His book, published a year later (Hope, *Household Furniture and Interior Decoration,* 1807), was a tribute to his innovative design and materially dictated the Regency style thereafter for neo-classical house furnishings: he was the first to institute the phrase "interior decoration" into the English language. He became known in Regency society as "*the costume and furniture man*", which he regarded as a compliment but which some critics, such as Lord Byron, deemed to be a term of ridicule. Following his death in 1831 at home in Duchess St, his son Henry Hope inherited

Fig. 7.17 Thomas Hope's Picture Gallery showing two pairs of classical Egyptian thrones with sphinx arms

the estate, but he sold the Duchess St family home in 1851 and it was demolished the same year. Deepdene also did not survive and was demolished in 1969, except for the mausoleum in the grounds.

7.4.2 The Tatham/Hope Egyptian Thrones

In a six day sale of The Hope Heirlooms between September 12th and 19th, 1917, Messrs. Humbert & Flint of Lincoln's Inn Fields advertised in Lot 1188 "*A Fine Pair of Lambeth Stone Chairs with square seats and with winged sphinx arms and a slab footrest",* there was another pair of similar chairs sold as a separate lot in Lot 1061. These were originally supplied to Thomas Hope at Duchess St and feature in drawings made of his Picture Gallery (Plate 2) first by Tatham and later by Westmacott (Westmacott, *British Galleries of Paintings and Sculpture: Comprising a General Historical and Critical Catalogue with Separate Notices of Every work of Fine Art in the Principal Collections,* 1824). A pair of these Sphinx Thrones from Duchess St./ Deepdene in Coade stone have recently appeared for sale with Edward Hurst Antiques of Shroton, Blandford Forum, Dorset, where they were described as follows:

A PAIR OF LATE GEORGE III COADE STONE THRONES: With framed slab back and plain slab seat, flanked by two winged sphinxes, their intricately carved wings forming the inner and outer arms of the throne, and the head, torso and feet of the sphinx forming the front uprights, with a shallow foot rest running between the two forward facing figures on either arm, both signed on the side of the footrest, one stamped 'Coade 1800', the other stamped 'Coade Lambeth 1800'. Designed by Thomas Hope (1769–1831), and produced by Coade in 1800. The finials were replaced by Coade Ltd. in 2021. w: 38″ (97 cm) d: 33″ (84 cm) h: 50″ (127 cm). Provenance: these thrones were designed by Thomas Hope for his house in Duchess Street.

These beautifully restored Coade stone artefacts (Figs. 7.18, 7.19, and 7.20), which are clearly marked and dated, can now be added to the small number of genuine Coade stone pieces that bear a dated mark (Table 1.3). Although they featured as a pair in the Humbert & Flint auction sale in 1917 they disappeared afterwards and have since been re-discovered. The second pair of Coade thrones from this sale are now at Parham House in Sussex; these are illustrated in Watkin and Hewat-Jaboor (Watkin & Hewat-Jaboor, *Thomas Hope, Regency Designer and Patron,* 2008, Fig. 4.16).

7.5 The King George III Statue at Lincoln

Sir Francis Dashwood (1708–1781), who formed a branch of the notorious Hell-Fire Club called the Order of the Friars of St Francis of Wycombe in the 1740s, married Sarah Ellys of Nocton, Lincolnshire, in 1745 and thereby acquired the estate for the Dunston Pillar, housing a lantern at the top which was lit each evening, as a "*beacon*

Fig. 7.18 Pair of Coade stone thrones with Sphinx arms designed for Thomas Hope's house in Duchess Street, London. (Reproduced with courtesy of Edward Hurst Antiques, Shroton, Blandford Forum, Dorset)

to guide the peasant, the wayfaring stranger and the horseman with his dame on pillion" across the lonely heath (Fig. 7.21, from the *Gentleman's Magazine*, July 1795, Plate III, page 561). The Dunston Pillar bore an inscription:

COLUMNAN HANC UTILITATI PUBLICAE
D.D.D. (Authors: Dono Dedit Dedicavit – Dedicated as a Gift)
F. DASHWOOD
MDCCLI

At the end of the eighteenth century, the Dunston Pillar was part of the estate that was inherited by Robert Hobart (1760–1816), the 4th Earl of Buckinghamshire, who removed the storm-damaged and then inoperational lantern and replaced it with a large Coade stone statue of King George III in his Garter robes, modelled by Joseph Panzetta and approximately 14 feet (~ 4.5 metres) in height, to celebrate his Golden Jubilee of 50 years on the throne, on 25th October 1809 (Fig. 7.22). The detail of the fur on the robes and the Garter chain moulded in Coade stone is still impressive. During the erection of the Coade stone statue on top of the Pillar, a stonemason, John

Fig. 7.19 Detail of Sphinx arm and rest on Coade stone throne shown in Fig. 7.17. (Reproduced with courtesy of Edward Hurst Antiques, Shroton, Blandford Forum, Dorset)

Fig. 7.20 Impressed date on Coade stone throne footrest, COADE 1800. (Reproduced with courtesy of Edward Hurst Antiques, Shroton, Blandford Forum, Dorset)

Wilson, fell to his death and he is buried in the nearby Harmston churchyard with an epitaph "*He who erected the noble king is here now laid dead by Death's sharp sting*".

In 1931, the statue was damaged in a storm and some pieces fell off and lay at the base of the Pillar: specifically the right forearm holding the sceptre. In 1941, it was necessary as a safety measure to reduce the height of the Dunston Pillar by about 30% to accommodate low-flying bombers returning from raids on Germany during

Fig. 7.21 The Dunston Pillar with its original lantern, Lincolnshire, from the *Gentlemen's Magazine*, July 1795, Plate III, p. 561

Fig. 7.22 The Dunston Pillar as modified in 1809 with the statue of King George III replacing the non-operational lantern

WWII to the Coleby Grange RAF airfield nearby: despite being instructed to take great care, the statue was badly fractured by the demolition men upon its removal and the King's head, for example, was found in twelve pieces and his torso became separated from his legs. The demolition process was likened at the time to Humpty Dumpty and the analogy was made of "putting King George together again". The surviving pieces of the statue, said to be many dozen in number, were housed in the Museum of Lincolnshire Life from between 1956 and 1970. The bust was restored in 1974 using a cementitious mortar which did not assist in its preservation because of an incompatible expansion coefficient with the Coade stone and a final further restoration was accomplished in 2010 by Coade Ltd., who specialised in the conservation of Coade stone artefacts: the bust was then mounted on a plinth in the grounds of Lincoln Castle (Fig. 7.23). This was considerably reduced in size from the original statue, representing its upper third only, and was affectionately then known as "George One-Third", while the rest of the statue fragments and pieces, including both legs, cloak and thighs, have remained stored in several sacks in a storeroom at Lincoln Castle (Folly Flaneuse 2020). The extent of the damage to the statue can be gauged from the remaining fragments, which comprise 45 large pieces plus 18 sacks of smaller pieces, numbering about 250 pieces in all. The larger pieces

Fig. 7.23 The George III bust in Lincoln Castle reassembled from fragments from the damaged statue that once sat atop the Dunston Pillar, 1810

show evidence of the procedure used by Eleanor Coade to assemble this very large statue using bronze dowels and a molten lead cast sealant, several of which have been replaced using iron rods, which means that the pieces of the statue were moulded, assembled as paste and then the statue was fired in the kiln (Skerritt 2006; Cowan 2008). Significant damage has been caused to the statue fragments since its demolition that have been attributed to poor storage and further weathering, and the overall condition of the surviving fragments has been judged to be very poor indeed. The use of metal rods in the assembly of the larger statues is interesting as this was not commented upon by Alison Kelly in her book (*Mrs Coade's Stone*, 1990) but was acknowledged in her later communication (Kelly 1996). The even larger statue of Lord Hill in Shrewsbury (height 5.2 metres), it is now realised also was fired from 12 pieces that were originally cramped together *in situ* atop its column using bronze armatures.

An interesting result from the first analytical investigation of the King George III statue at Lincoln by Karran and Colston in 2016 (Karran and Colston 2016) is that the determination of the elemental oxide percentages by the XRF technique was inconclusive for the reconstructed head but confirmed that the legs which were held separately in store in Lincoln Castle were definitively made of Coade stone. The authors of this paper were of the opinion that the materials used for assembly of the shattered head and bust compromised the recording of the elemental oxide composition and this needs further study. Because some if not many Coade stone artefacts have at some time been restored, conserved or cleaned then the ability of the various analytical techniques to determine their composition needs to be evaluated comprehensively: it could be that all the analytical techniques used are so compromised or perhaps only specific techniques are affected?

In our own analyses, some four specimens were carefully selected by the Curator from the larger pieces in the sacks in store, that were not compromised by the weathered iron rods and ties and by a resultant delamination: these were taken from the cloak, thigh and legs of the statue and also included a specimen from the interior artificial stone. The analyses confirm that the statue fragments are indeed composed of Coade stone.

7.6 St George's Chapel, Windsor Castle

One of the most magnificent settings for Eleanor Coade's artificial stone artefacts was St George's Chapel, Windsor Castle. A late mediaeval building in the Perpendicular Gothic style, it was constructed in the thirteenth century by King Henry III as a chapel to St Edward the Confessor and was enlarged by King Edward III in 1348. It is a Royal Peculiar, namely a church that is under the direct jurisdiction of the Monarch, and also serves as a Chapel of the Knights of the Order of the Garter. Between 1787 and 1790, King George III personally commissioned the architect Henry Emlyn to make alterations to St George's Chapel: the original screen between the nave and the chancel quire was removed and a Coade stone replicate was

Fig. 7.24 St George's Chapel, Windsor Castle: fan vaulting in Coade stone, *ca.* 1799

substituted (Fig. 7.24). This comprised three four-centred Tudor arches with Gothic ornaments on spandrels and bosses of St George and the Dragon, roses and a monogram of King George III inside the Order of the Garter. Inside the replacement screen the fan vaulted ceiling emerges from thin columns, all manufactured in Coade stone to match the late fifteenth century vaulting in the Quire aisles. Additionally, there is a Coade stone frieze on the King Edward IV monument and a decorative niche in the North Quire Aisle. There are also three statues in niches on the western front of the Chapel representing St Edward the Confessor, St George and the Virgin Mary that were produced by Eleanor Coade in Coade stone a while later in 1799. A Coade stone Gothic font made from a standard stock item was also ordered by Canon Dr. Majendie and delivered but this was destroyed by Queen Victoria in 1879 to make way for her statue of the Prince Imperial, son and heir of King Napoleon III of France, whose foolhardy and reckless pursuit against superior orders of a retreating Zulu impi in Natal whilst he was serving there with a British cavalry regiment in the Second Zulu War of 1879–1880 resulted in his encirclement, entrapment and death. A further replacement was undertaken in the 1920s when part of the front of the Coade stone screen was replaced by natural stone because of a "*misunderstanding of what the Coade stone comprised*" (Darlington 1955).

This major Royal commission must have heartened Eleanor Coade immensely and it has been commented that most of the several hundreds of thousands of visitors to St George's Chapel annually do not realise that what they are seeing there in the Nave and Chancel Quire is not mediaeval fourteenth century natural stone at all but rather an eighteenth century artificial Coade stone ceramic replacement—or as Alison Kelly (Kelly, *Mrs Coade's Stone*, 1990, p. 217) put it rather succinctly "*a vast piece of pottery*".

7.7 Bloxholm Manor and Bloxholm Church

In 1632 the manor of Bloxholm passed to Nathaniel Hubbard and thence to Cyprian Thornton, who built the Hall in the mid-to-late seventeenth century. The estate was purchased by the Duchess of Rutland and passed to her grandson, Robert, General Lord Robert Manners, Marquis of Granby, who enlarged the Hall and built a central gable feature whereon was mounted his coat-of-arms and a peacock in Coade stone in 1772 (Fig. 7.25). The Hall was abandoned in the late 1940s and was partially demolished in 1963; the Manners coat-of-arms in Coade stone is now found above the entrance to the Church of St Mary the Virgin but the very fine peacock, representing the Manners crest is lost. It seems that Robert Manners was a member of the elite Hell-Fire Club, also known as the "Lincoln Club" locally, which was headed by Sir Francis Dashwood Bt, who erected the Dunston Pillar referred to above in another case study (*viz.* the statue of King George III).

Fig. 7.25 Coat of arms and crest of General Lord Robert Manners, Bloxholm Hall, *ca.* 1800, now relocated to Bloxholm Church

7.8 Anglesey Abbey

The statues of caryatids (Fig. 7.26) that are located in the gardens of Anglesey Abbey, situated some 8 miles from Cambridge, came from Buckingham House, Pall Mall, London, not to be confused with Buckingham Palace situated at one end of The Mall. Between 1792 and 1795, Buckingham House was redesigned by Sir John Soane as the London residence of the Duke of Buckingham and Chandos as a neo-Palladian mansion (Fig. 7.27), who first used there eight caryatids in support of an oval lantern above the staircase in what became his favourite classical statue in Coade stone. Sir John Soane possessed his own personal copy of Eleanor Coade's Catalogue of 1784 with its 784 classical designs, which is now preserved in the Sir John Soane Museum, London. He later used these groups of caryatids in other prestigious projects, such as twelve more for support of the rotunda dome in the Bank of England and for his own country house at Pitzhanger Manor and his London town house at 13, Lincoln's Inn Fields. It is believed that these caryatids were designed and sculpted for Eleanor Coade originally in 1790 by John Rossi, or perhaps John Bacon, modelled upon the sanctuary of Pandrosius in Athens and

Fig. 7.26 Caryatid from the gardens at Anglesey Abbey, Cambridgeshire

Fig. 7.27 Buckingham House, Pall Mall

their importance in the Coade repertoire was that they were available for manufacture many years before Lord Elgin rescued the caryatid statues from the Erychtheion in Athens in 1816. Buckingham House was demolished in 1908 and at the closing-down sale the caryatids were purchased by Lord Fairhaven and relocated to his estate at Anglesey Abbey in Cambridgeshire in 1926: the six surviving caryatids now form a Coronation Avenue in the gardens at Anglesey Abbey. Only three are marked and impressed: with COADE LAMBETH 1793, COADE LONDON and COADE LAMBETH (Roper, *The Gardens of Anglesey Abbey, Cambridgeshire*, 1964). Caryatids designed by Rossi were also used to support a roof structure at St Pancras' Church, London, where a vertical measurement error by the architect required the post-firing removal of a section of the statues near the midriff and their subsequent rejoining together to make them fit, which can still be observed today (Fig. 7.28).

It appears that the caryatids in Buckingham House originally were painted to simulate bronze sculptures as there are still black discoloured residues adhering to their surface: as Karran and Colston (2016) have noted during their analytical studies, this compromises the analytical determination of the elemental oxide percentages using non-destructive HHXRF and affects their interpretation of the Coade stone composition.

Fig. 7.28 St Pancras Church, London—four caryatids in Coade stone

7.9 Coade Stone Artefacts at Farnley Hall, Otley, Yorkshire

Farnley Hall is an imposing eighteenth century mansion near Otley, North Yorkshire, and was once the home of Walter Ramsden Hawksworth Fawkes, who was an MP and High Sheriff for Yorkshire during the early part of the nineteenth century and a dedicated patron of the fine arts. A lifelong friend of the artist James Mallord William Turner, he once possessed the finest collection of Turner oil paintings and watercolours that were amassed during Turner's holidays and painting sojourns at Farnley Hall between 1805 and 1825. Alison Kelly's book (*Mrs Coade's Stone*, 1990) mentions several items of Coade stone that exist at Farnley Hall, which can be summarised as follows:

- July 1818: "Mr Fawkes, fountain from Mr Rhodes' drawing, 40 gns" (CO 1818). Taken from William Croggon's Order Book that is filed with Croggon's Day Book of accounts at the Public Records Office (C111/106). In the eighteenth and early nineteenth centuries fountains were characteristic features of formal gardens in the Italianate style and, although they fell out of favour in the late eighteenth century, they were revived in the early nineteenth century and became fashionable for the Regency period especially constructed as "monopodia". In Eleanor Coade's Gallery catalogue of 1799 an example was given of a fountain which was described as being modelled upon Bernini's Triton fountain in the Piazza

Barberini, Rome. None of these Coade Triton fountains have survived but a single example of a similar Coade stone fountain based upon a human figure after Giambologna is to be seen at Petworth House, Sussex. Alison Kelly believes that perhaps Eleanor Coade mistakenly attributed her fountain model to Bernini instead of Giambologna. Another example of a Coade stone fountain is known at Kearnsey Abbey, Dover, where it was supplied to a Mr. Fector after a design by Henry Rhodes in September 1818 and appears in the Coade ledgers as being supplied at a cost of 35 gns: the artificial stone design was based on a coral tripod stand bearing a stone fountain and coral top supported by shells. Henry Rhodes (d. 1846), who was an architect involved in both Farnley Hall and at Kearnsey Abbey, exhibited his work "*for a nobleman's mansion*" at the Royal Academy in the early 1800s (Colvin, *A Biographical Directory of British Architects, 1600–1840*, 1978, pp. 808–809). Colvin's work cites 2000 architects from Inigo Jones (1573–1652) to Sir Charles Barry (1795–1860). William Croggon also lists orders for 13 fountains in his Day Book, but these comprise mainly models which utilise shells and dolphins, which hence makes the Farnley Hall fountain a unique survivor; the Farnley Hall Coade stone fountain is unique in its design which incorporates "*Three basons with lion's heads supported by lion's legs, £42 (40 gns.)*".

- The Farnley Hall entry in Croggon's Order Book mentioned above also states "July 1818: Walter Fawkes Esq., Farnley Hall, Otley, Yorkshire, two hawks for gate piers, £29 8s 0d". Whereas birds are quite commonly represented in Coade stone moulds for garden statuary, some usual examples being peacocks, eagles and swans, the two hawks at Farnley Hall are again unique survivors of their type, which Alison Kelly notes were in good condition still a few years ago. The hawks situated atop the gate piers at Farnley Hall were painted by J.M.W. Turner in his watercolour of the house, which still remains there (Fig. 7.29): Turner first visited Farnley Hall in 1796 from a stay at Harewood House nearby, where he painted for Edward Lascelles, Baron Harewood and later Viscount Lascelles, and continued his painting visits to Yorkshire and to Farnley Hall until his host, Walter Fawkes, died in 1825.

The third example of Coade stone at Farnley Hall that is recorded in Croggon's Order Book comprises "two shields of arms, "£29 8s 9d". Walter Ramsden Hawksworth Fawkes was born on the 2nd March 1769 in Otley and he died on October 24th 1825, aged 56. He was the son of Walter Ramsden Beaumont Fawkes and Amelia Farrer. In 1794 he married Maria, daughter of Robert Grimston of Neswick Hall and they had four sons and seven daughters; Maria died in 1813 and Walter then married Maria Sophia, daughter of John Vernon of Clontarf Castle and the widow of the Rev. the Hon. Pierce Butler (1782–1808), third son of Henry Butler, 2nd Earl of Carrick, in 1816. Walter Fawkes completed the Grand Tour in 1794 and became MP as a Whig for Yorkshire in 1806 and a High Sheriff in 1823 (R. Thorne, *The History of Parliament: 1790–1820*, 1986). He raised the Wharfedale Volunteers against Napoleon Buonaparte in 1806. Fairbairn (Fairbairn, *Crests of the Families of Great Britain and Ireland and Their Kindred in Other Lands,* 1905 and

Fig. 7.29 Detail from a watercolour of Farnley Hall, East Lodge gateway, showing the two Coade Stone falcons atop their gatehouse piers facing the carriageway, painted by James William Mallord Turner, *ca.* 1820. (*Courtesy of Guy Horton-Fawkes Esq., Farnley Hall, Otley*)

1986) gives the crest of Walter Fawkes as a falcon proper and his motto as *A Deo et Rege,* which translates as *From God and the King*. Sir Bernard Burke (Burke, *Burke's General Armory—The General Armory of England, Scotland, Ireland and Wales: Comprising a Registry of Armorial Bearings from the Earliest to the Present Time*, 1884) gives the coat of arms for Walter Fawkes in his General Armory as, ermine a mascle sable, which has an escutcheon of ermine on which is displayed a black lined voided lozenge, known as a mascle heraldically, centrally and vertically (Cussans, *The Grammar of Heraldry*, 1866). No trace of these coats of arms paterae now remain at Farnley Hall. However, on the cover of Finberg's book (Finberg, *Turner's Watercolours at Farnley Hall*, 1912) the arms and crest of Walter Ramsden Hawksworth Fawkes appears and this detail is shown in Fig. 7.30, in which the heraldic ermine field with the mascle are shown.

Through the kindness of Guy Horton-Fawkes Esq., the current incumbent of Farnley Hall and direct descendant of Walter Ramsden Hawksworth Fawkes, one of the authors was permitted to examine the Coade stone artefacts at Farnley Hall. The hawks that were originally situated atop the gatehouse piers were unfortunately vandalised during an attempted theft some years ago but the fragments have been retained in storage and specimens could now be taken for analysis. Their provenance therefore gives a compositional baseline for Coade stone that was manufactured in 1818. A photograph of J.M.W. Turner's watercolour, which still hangs in Farnley Hall, is shown here also (Fig. 7.29) and depicts the two hawks in their original position atop the gatehouse piers facing inwards towards the carriageway. As

Fig. 7.30 Detail of the coat of arms of Walter Ramsden Hawksworth Fawkes as it appears on the cover of Finsberg's book, *Turner's Watercolours at Farnley Hall, 1912*

mentioned above, the coat of arms of Walter Ramsden Hawksworth Fawkes has a falcon crest and the two Coade stone falcon artefacts must therefore have been modelled upon this. A photograph of the East Lodge gateway to Farnley Hall is shown in Fig. 7.31, with a pier some 20 feet in height. The two falcons which originally sat atop these gatehouse piers were fragmented in their attempted theft: one has been repaired (Fig. 7.32) and the other is rather more broken (Fig. 7.33). The latter clearly displays a mascle (a voided lozenge heraldically) upon its breast as shown in Fig. 7.34—the other also displays a mascle but it is not so clearly depicted. The presence of the mascle, which features prominently on the Fawkes escutcheon, renders the Coade stone falcon figures personal to Walter Ramsden Hawksworth Fawkes. The better preserved falcon measures some 75 cm in height and sits on a circular Coade stone plinth of diameter 60 cm. The broken falcon afforded an excellent opportunity to sample both the interior and exterior surfaces of Coade stone for analysis.

Finally, the lion monopodia fountain whose original location at Farnley Hall is uncertain, is also mentioned in William Croggon's account books and itself provides yet another unique example manufactured in Coade stone. Currently, the fountain is also damaged and fragmented but the three lion's heads are still clearly visible supporting the bowl of diameter 1 metre and depth 40 cm (Figs. 7.35 and 7.36), and the plinth upon which it once stood is also present but now detached (Fig. 7.37).

Fig. 7.31 Photograph of the East Lodge gateway, Farnley Hall, showing the left hand stone pier, now devoid of its falcon statue which was damaged in an attempted theft some years ago. (*Courtesy of Guy Horton-Fawkes Esq., Farnley Hall, Otley*)

7.10 Stourhead, Stourton-With -Gasper, Mere, Wiltshire

Stourhead is a Palladian mansion which was built on the site of an older residence by Nathaniel Ireson and Colen Campbell between 1721 and 1725 for Henry Hoare, a wealthy banker, who purchased it from the Stourton family in 1717. It was destroyed by fire and rebuilt in 1906. The Temple of Flora is situated in the grounds near a large lake formed by damming a stream in a landscape designed by Henry Hoare. In the Temple of Flora is a large Coade stone vase modelled upon the Borghese Vase, which with the Medici Vase was one of the two most admired classical vases from antiquity which were seen by travellers on the Grand Tour in the eighteenth century and the originals could be seen at the Villa Borghese and the Villa Medici in Rome at that time. The original Borghese Vase is now in the Louvre Museum, Paris, where it was deposited by Napoleon, who purchased it from Camillo Borghese in 1808 (Fig. 7.38). The original Medici vase, a smaller version, was for some time in the Villa Medici but is now in the Uffizi Gallery, Florence (Haskell and Penny, *Taste and the Antique: The Lure of Classical Sculpture, 1500–1900*, 1981). Eleanor Coade marketed her copies of the Borghese and Medici Vases (Nos. 81 and 82 respectively in her Catalogue of 1784), which were very popular garden ornaments and could be

Fig. 7.32 Coade stone falcon, damaged, from the gatehouse pier in storage at Farnley Hall. (*Courtesy of Guy Horton-Fawkes Esq., Farnley Hall, Otley*)

Fig. 7.33 The other more seriously damaged Coade stone falcon, the pair to that shown in Fig. 7.32, showing its hollowed interior. (*Courtesy of Guy Horton-Fawkes Esq., Farnley Hall, Otley*)

Fig. 7.34 Detail of the heraldic mascle device on the breast of the falcon shown in Fig. 7.33, which personalises it to Walter Ramsden Hawksworth Fawkes from his coat- of-arms. (*Courtesy of Guy Horton-Fawkes Esq., Farnley Hall, Otley*)

purchased as a pair or singly—the Coade stone Medici Vase is shown in Fig. 7.39. The artistic decoration of the Coade stone Borghese and Medici Vases will be described in detail in the next case study for Kedleston Hall, where they exist as a pair situated on the *piano nobile* on either side of the front door.

The Stourhead Borghese Vase is shown in Fig. 7.40 and it is on record that it was purchased by Henry Hoare for location in the Temple of Flora in his gardens for £12 14 s 6d, and was delivered in 1770 by Eleanor Coade, but from an original order placed with Daniel Pincot in 1769 for an initial fee of £5 5s 0d (Caroline Stanford, personal communication). It has certainly been subjected to some restoration and therefore a careful sampling excision for analysis as reported in Chap. 6 was made in the region of one of the satyr's heads (Fig. 7.41).

7.11 Kedleston Hall, Derbyshire

Coade stone Borghese and Medici Vases for Kedleston Hall were purchased from Eleanor Coade by Nathaniel Curzon, 1st Baron Scarsdale, in 1771 and presumably are exemplars of the earliest pair of these vases that are still in existence. They are

Fig. 7.35 Fountain bowl in storage at Farnley Hall, damaged, showing the three lion's head supporters arranged symmetrically around the rim. (*Courtesy of Guy Horton-Fawkes Esq., Farnley Hall, Otley*)

Fig. 7.36 Lion's head detail on the Coade stone fountain bowl shown in Fig. 7.35. (*Courtesy of Guy Horton-Fawkes Esq., Farnley Hall, Otley*)

still housed at Kedleston Hall and are situated in their original positions flanking the entrance to the house from the *piano nobile*. They now form part of the Scarsdale Collection acquired with the assistance of the National Heritage Memorial Fund and were transferred to the National Trust in 1987.

Fig. 7.37 Coade stone plinth to the base of the fountain bowl shown in Fig. 7.35. (*Courtesy of Guy Horton-Fawkes Esq., Farnley Hall, Otley*)

Fig. 7.38 The Borghese Vase, The Louvre Museum, Paris. (*Creative Commons CCO. Public domain*)

Fig. 7.39 The Medici Vase, Coade stone copy, the Royal Gardens Kew. (*Creative Commons CCO. Public domain*)

Fig. 7.40 The Coade stone Borghese Vase in the Temple of Flora at Stourhead, Wiltshire. (*Courtesy of Caroline Stanford*)

Fig. 7.41 The sampling region of the Borghese Vase at Stourhead near the satyr's head. (*Courtesy of Caroline Stanford*)

7.11.1 The Borghese Vase

Both the original Borghese and Medici Vases are bell shaped krater sculpted from Pentellic marble in Athens sometime during the first century BCE, although some historians feel that the Medici Vase is slightly later and was made specifically for a Roman audience: the Borghese Vase is now in the Louvre Museum, Paris and the Medici Vase is in the Uffizi Gallery, Florence. Eleanor Coade made her copies in Coade stone of these vases from about 1770 until 1820 and they were very popular items of garden statuary and ornamentation. The Borghese Vase, 1.72 m high and 1.35 m diameter, is moulded with a vine motif and has a bas relief depicting the thiasus, a Bacchanalian procession presided over by Dionysius and Ariadne. The lower section is gadrooned with a pair of mascarons shaped in the form of satyr's heads. The Borghese Vase is mounted on a tapering fluted stem and a low octagonal plinth which was usually mounted on a sandstone plinth. It is stamped COADE, LONDON.

7.11.2 The Medici Vase

The Medici Vase is slightly smaller than the Borghese Vase and measures 1.52 m in height with a diameter of approximately 1.25 m. Again, moulded with a vine leaf motif the bas relief now tells the story of Iphigenia seated behind Diana and her warriors (Agamemnon, Achilles and Odysseus). It differs from the Borghese Vase in that it has two fluted loop handles which rise from paired satyr's heads on each side and has acanthus leaves on the spreading gadrooned base located on a small square plinth. It is stamped COADE, LONDON. Fecit.

7.12 Castle Howard, Yorkshire

The Earl of Carlisle ordered a Coade stone copy of the Medici Vase from Eleanor Coade on the 19th February 1778 and this is shown in Fig. 7.42. Unlike the original ancient marble vases which are preserved in museums, the Coade stone replicas are sometimes found outdoors in landscaped gardens and the Castle Howard version is

Fig. 7.42 The Coade stone Medici vase on its sandstone plinth at Castle Howard, Yorkshire. (*Courtesy of Eleanor Brooke-Peat, Curator of Collections at Castle Howard*)

no exception: it is mounted on a square sectioned sandstone plinth which displays paterae of Graeco-Roman gods.

7.13 Wentworth Woodhouse, Rotherham, Yorkshire

A Jacobean mansion founded by Thomas Wentworth, the 1st Earl of Strafford under King Charles I, was rebuilt by Thomas Watson-Wentworth between 1725 and 1728, the 1st Marquess of Rockingham (1693–1750). After Charles Watson-Wentworth, the 2nd Marquess of Rockingham, the estate passed through inheritance to the Marquess' sister and the Earls Fitzwilliam who owned it until 1979, when it fell into disrepair. It is a very grand mansion with an imposing East Front (Fig. 7.43). Several artificial stone artefacts were ordered by Lord Rockingham, the earliest being commissioned from Daniel Pincot in 1767, and thereafter from Eleanor Coade from the 1780s through to 1814, However, when the estate became derelict in the late twentieth century, the contents were sold off. Currently, the house and estate are being refurbished and several artificial stone items have been reacquired; two of these are the Dying Gladiator, also known as the Dying Gaul or Dying Galatian (*Galata Morenta*), which sit now to the right and the left at the base of the main staircase leading up from the Entrance Hall (Figs. 7.44 and 7.45).

Fig. 7.43 The East Front of Wentworth Woodhouse, Rotherham, Yorkshire. (*Image courtesy of Andrew Abbott, CC by SA 4.0*)

Fig. 7.44 The Dying Gaul statue in the entrance hall of Wentworth Woodhouse, situated to the left of the staircase. (*Courtesy of Caroline Stanford*)

Fig. 7.45 The Dying Gaul statue in the entrance hall of Wentworth Woodhouse, one of pair with that shown in Fig. 7.44, situated to the right of the staircase. (*Courtesy of Caroline Stanford*)

7.13.1 The Dying Gladiator

An ancient Roman marble statue, now in the Capitoline Museum, Rome, was a copy of a lost bronze sculpture from the Hellenistic period (323–31 BCE) that was originally made, and attributed to Epigouris, to celebrate the victory of Attalus of Pergamon, *ca*. 235 BCE, over the Celts (Galatians) in Anatolia. The semi-recumbent figure shows a fatally wounded Celtic soldier with a sword wound in his chest and wearing the characteristic Celtic torc. It was first recorded in 1623 in the Ludovisi family collection and was then relocated to the Capitoline Museum, from where it was liberated by Napoleon and after spending some time in the Louvre was later repatriated to Rome in 1814. It was certainly one of the most celebrated classical sculptures of antiquity to be viewed on the Grand Tour in the eighteenth century and featured prominently on the list of "must be seen". Lord Byron was so impressed with its pathos that he wrote it into his poem *Childe Harold's Pilgrimage* (4th canto, 1818). An early example in England was made in bronze for the Duke of Northumberland at Syon House in 1773 and another in bronze is at Iford Manor and has been illustrated in Fig. AII.4.

7.14 Harlaxton Manor, Grantham, Lincolnshire/Witley Court, Worcestershire

The Harlaxton Manor house (Fig. 7.46) was constructed in 1831 by the architects Anthony Salvin and William Burn for Nottingham businessman Gregory Gregory in a mixture of Jacobean, Gothic and baroque architectural styles with Italianate and Dutch landscaped gardens, an a site formerly listed as Harleston Manor which had been developed by Sir Daniel de Ligne in 1619 that was inherited by Gregory Gregory from his uncle, George de Ligne Gregory (1740–1822). Violet van der

Fig. 7.46 Harlaxton Manor, Grantham, Lincolnshire: the forecourt, showing the two Coade stone lions holding heraldic escutcheons and flanking the stone steps. (*Courtesy of Richard Croft, CC by SA 2.0*)

Fig. 7.47 Witley Court, Worcestershire: a print dating from 1880

Elst ran the estate, renaming it Grantham Castle temporarily, in the twentieth century and acquired the Coade stone Harlaxton lions from Witley Court following a disastrous fire there in 1937.

Witley Court (Fig. 7.47) was a Jacobean manor house, built in 1655 and then sold to the Foley family who occupied it until it was sold to Lord Ward, later the 1st Earl of Dudley, in 1837. At the turn of the eighteenth century, Lord Thomas Foley commissioned John Nash to extensively develop Witley Court into a very grand mansion. The Earl of Dudley leased Witley Court to Queen Alexandra, widow of King William IV, for several years from 1837 and revamped the house and estate from 1846 into the 1850s and then the 2nd Earl made some more extensive developments from 1888 until the estate was eventually sold off to a local industrialist, Sir Herbert Smith, in 1920. In 1937 a disastrous fire gutted a large part of the building and because it had been unoccupied it was not covered by insurance, so the surviving contents and garden statuary were sold off at auction. Baron Foley of Kidderminster (motto, *Ut prosim*, meaning *That I May Serve*) had commissioned Eleanor Coade to provide two lions for Witley Court: these were assumed based upon his crest, which is a rampant lion bearing his coat of arms, an escutcheon charged with a fess engrailed between three cinquefoil, all within a bordure sable. The Harlaxton lions are now placed on plinths adjoining steps on the forecourt (Fig. 7.46) of Harlaxton Manor and a detail shows one rampant lion holding a coat of arms between his forepaws (Fig. 7.48) and an accompanying impressed stamp and date, COADE & SEALY 1811(Fig. 7.49). However, close inspection reveals that the escutcheon does not represent the Foley arms but rather those of Lord Ward, namely a shield chequy of gold and blue with a bend ermine with the appropriate hatchings

Fig. 7.48 The Harlaxton Manor/Witley Court lion, bearing the heraldic escutcheon of Lord Foley between its forepaws

Fig. 7.49 Impressed and dated Coade stone stamp on Harlaxton/Witley Court lion, COADE & SEALY 1811, as ordered by Baron Thomas Foley of Witley Court, Worcestershire

for blue and gold (respectively horizontal bars and dots). Chronologically, therefore, the arms of Ward would not have been a part of the original Coade statue and are a later replacement, probably in the 1850s; the commutability of the escutcheon would have been a desirable attribute of the Coade stone statue as it would facilitate its survival and acquisition by other armigerous families. It is likely, therefore that the replacement escutcheon would be shown upon analysis to be not of an original Coade stone composition.

In 1971, the University of Evansville, Indiana, USA, a private University endowed in 1854, purchased Harlaxton Manor as its British campus, where it now holds courses in Europe for international students.

7.15 The Sundial Plinth at Weston Park, Shifnal, Shropshire

There is an invoice held at Weston Park, Shifnal, The Wrekin, Shropshire, dated January 1st, 1779, for a three -sided sundial plinth in Coade stone to be supplied from Eleanor Coade in Lambeth for the sum of £28-0-0 to Sir Henry and Lady Bridgeman. The sundial plinth was last seen in a photograph from the early twentieth century showing Margaret, Countess of Bradford, and her children standing in the formal gardens laid out by the 3rd Earl of Bradford, dating perhaps from the late 1910s. The 2nd Earl of Bradford's title became extinct in the early eighteenth century and the estate passed to his daughter, Lady Anne Newport, who married Sir Orlando Bridgeman Bt. Their son, Sir Henry Bridgeman Bt, lived at Weston Park from the 1760s and he made extensive alterations to the house and engaged Lancelot "Capability" Brown to landscape the gardens. Sir Henry was elevated to the peerage as Baron Bradford and later the Earldom of Bradford was reinstituted. This Coade stone artefact was apparently then lost for many years until it was discovered in a sad condition by the Head Gardener, Martin Gee, and his team at Weston Park buried beneath a mound of garden rubble, detritus and overgrown vegetation in 2017. Restoration of the sundial plinth has now been achieved and a new brass sundial gnomon has been crafted by Robert Foster of Jackfield. The Coade stone sundial and its plinth, after its restoration by Philip Thomason of Cudworth, can now be seen as fresh and beautiful as it was first made in the eighteenth century: Gareth Williams, Curator at Weston Park, discovered the original invoice for the sundial plinth in the house accounts archive and he has recently produced a book (Williams, *Weston Park: The House, the Families and The Influence,* 2022). Weston Park is believed to be the real location of Blandings Castle, the setting for a series of novels by P.G. Wodehouse and featuring his famous butler, Jeeves, and Lord Bertie Wooster.

The discovery of this sundial plinth appears in the known list of existing Coade stone exemplars that was formulated by Alison Kelly over the twenty years or so of her comprehensive research investigation (Kelly, *Mrs Coade's Stone,* 1990) but she refers to it third hand as information received from a local historian, John Duxfield, and perhaps it was no longer visible when she was compiling her survey of Coade stone artefacts. She generally makes reference to local historical accounts when the artefact was no longer visibly extant but, of course, it is quite possible that these sculptures were no longer at their designated locations. The sundial and plinth are mentioned in the Coade & Sealy Catalogue of 1784, as item 144 and also item 533, both 4 feet high, where they are costed in at the sums of £4-4-0 and £18-18-0, respectively, namely 4 guineas and 18 guineas. These prices are both significantly less than the sum Sir Henry Bridgeman paid in his invoice from Eleanor Coade, so it can be assumed that perhaps some additional embellishment was required to the standard moulding for the Weston Park sundial plinth or perhaps some additional mounting was required and, of course, there would have been significant transportation costs incurred too for its carriage from Lambeth to Shropshire. As far as can be ascertained, there is no mark on the sundial plinth, which is a great pity because otherwise this would have represented one of the earliest pieces of marked Coade stone supported with documentary historical support. Further examples of Coade stone sundials of a similar design and tripod type with supporting caryatids to that seen at Weston Park are known at Audley End, Wardour Castle and Syon House and a recently restored exemplar sold at auction for £55,000.

An interesting corollary emerges in a note that the Weston Park estate has apparently a second sundial in Coade stone which is located at the edge of Capability Brown's landscape development at Temple Wood. Although this does not appear in the house archival records, it does bear the stamp of Austin & Seeley upon the brass sundial itself, which immediately raises the question as to whether or not it is truly a Coade stone artefact or perhaps is best categorised as a cementitious replica; it has already been pointed out that Felix Austin, along with other artificial stone manufacturers, possibly purchased several moulds from the winding-up sale of Eleanor Coade's business which was then in the hands of Thomas Croggon up to 1841 and that he used these to reproduce Coade statuary in cementitious stone reinforced with iron struts. Two possibilities are, therefore, that this second sundial plinth is not Coade stone at all but rather a cement composite or alternatively, that Austin & Seeley provided a replacement sundial fitment for the top of an existing and earlier Coade stone plinth. Clearly, an analytical study of the composition of the sundial and its plinth would establish which of these alternatives is the correct supposition.

7.16 The Nelson Memorial, Royal Naval College, Greenwich

The Nelson Memorial at the Royal Naval College, Greenwich, was commissioned from Eleanor Coade in 1810–1812 and is undoubtedly the largest sculpture made in Coade stone, being some 40 feet in overall length (Fig. 7.50) and 10 feet in height, being encased by the pediment installed facing a quadrangle in the King William III Block by Sir Christopher Wren. It was sculpted by Joseph Panzetta to a design of Benjamin West and displays the body of Nelson being offered by Neptune to a helmeted Britannia flanked by angels, a shield and a lion. Although a plaque states CXXII BATTLES, only Copenhagen, The Nile and Trafalgar are specifically mentioned: it is not possible that Nelson fought 122 battles as stated here, although this may include what in naval terms were classed as "cutting-out engagements" in which small boats or cutters were used by naval detachments to board enemy vessels

Fig. 7.50 The Coade stone Nelson Memorial at the Royal Naval College, 1812, Greenwich, sculpted by Joseph Panzetta

during a blockade. It is well regarded as Eleanor Coade's finest work, it took three years to complete and cost the sum of £2584 at that time, which is equivalent to approximately £250, 000 today.

Eleanor Coade and John Sealy published a pamphlet on it entitled "A Description of the Grand Model of Neptune giving up the body of Nelson" (Coade and Sealy 1813) in which they state "*...The only proper medium to do justice to the lives of men who had signalized themselves in the service of their country – an Epic or Heroic Poem in Sculpture. Britannia is seated on a rock washed by the Ocean receiving the dead body of Nelson, given up to her arms at the command of Neptune by a Triton*". The British lion is shown with the flags of the three kingdoms of England, Scotland and Ireland.

7.17 Malvern Hall, Solihull, Warwickshire

Malvern Hall, Solihull, was built by Humphrey Greswolde in the early eighteenth century, whose descendant, Henry Greswolde Lewis, was a patron of John Constable. In 1818, Henry Greswolde Lewis ordered four Coade stone statues from Eleanor Coade at a cost of £157 8s 4d. These were classical figures representing Vesta, Sibyl, an Ancient Briton, and a Norman: the Vesta and Sibyl statues were placed in niches in the gatehouse designed by Sir John Soane and the Ancient Briton and Norman statues were placed on plinths to the left and right, respectively, of the house entrance as shown in the Constable painting of 1821 in Fig. 7.51, which also shows on the far right the pair of gatehouse piers which housed the figures of Vestal and Sibyl. Vestal and Sybil were offered as standard figures in Eleanor Coade's first

Fig. 7.51 Malvern Hall by John Constable, oil on canvas 54 × 78 cm, signed and dated 1821, showing the Coade stone figures of an Ancient Briton and a Norman on their plinths either side of the entrance to the house. (*Courtesy of the Clark Art Institute, Williamstown, Massachusetts, USA, accession 1955.683*)

Catalogue of 1784, numbered 4 and 5 respectively, and came in three sizes, the largest at 4′6″ (1. 37 m) and costing 16 gns each, but the Ancient Briton and Norman statues do not feature in this Catalogue. This painting was commissioned by Magdalene Lewis, the Dowager Countess of Dysart (died 1823), sister of Henry Greswolde Lewis, who was also a patron of John Constable. Constable painted several versions of Malvern Hall and an earlier one dated 1809 does not show the Coade stone figures on their plinths in front of the house. A more recent photograph from the 1920s (Fig. 7.52) shows the house entrance with these two Coade stone figures of the Ancient Briton and the Norman on their plinths. Figure 7.53 shows the figures of Vesta and Sibyl in their gatehouse niches.

The estate was wound up and the contents of Malvern Hall were sold at auction in 1922: thereafter it became Solihull Girls' High School, whereupon the headmistress was charged by the Board of Governors with locating and replacing the missing figures from the auction sale, which was duly accomplished. The staff and students immediately adopted the figures and cherished them: they featured on special

Fig. 7.52 Malvern Hall, Solihull, photograph *ca.* 1920, showing the Coade stone figures of an Ancient Briton and a Norman in place approximately two years before the estate was wound up and sold at auction. (*Courtesy of Tracey Williams, The Core, Solihull, UK*)

Fig. 7.53 Coade stone figures of Vesta and Sibyl in their niches in the gatehouse piers designed by Sir John Soane. (*Coutrtesy of Tracey Williams, The Core, Solihull, UK*)

occasions in the school calendar and a school poem was written focussing on Vestal and Sibyl:

"They sought those statues for eighteen years
A most inspiring tale.
Until at last through Fortune's whim
They found them at a sale.
"Bid anything" the Governors said,
"But buy them without fail".
"Ah ! Got them " the Headmistress said,
"I've always had an itch
To get that pair of statues back.
I hate an empty niche!"

The other two statues of the Ancient Briton and the Norman were christened Alfred and Sigismund, but an unfortunate accident occurred in the early 1950s when a tradesman's vehicle reversed into the plinth holding the statue of Sigismund, the Norman, which fell off its plinth and was shattered. This resulted in another school poem being written, highlighting the loneliness of Alfred, the Ancient Briton:

Alfred's Lament (authors: apologies to John Masefield!)
"I must go back to my plinth again
To a plinth with a concrete base.
All I want is a cosy niche
Instead of a packing case
A lofty perch, a commanding view, a landscape fore and aft,
Exposed to elements great and strong.
Instead of a nagging draught.
I remember Sig, my friend, for Sig was here for ages, destroyed in his prime.
I was very sad and very touched when Sig was struck down by a lorry.
All is well when summer's here but I am in a windy spot right now ...."

This highlights the fragility of Coade stone sculptures towards mechanical shock, resistant though they obviously are to weathering exposure under normal environmental conditions.

7.18 Lord Schomberg's House, Pall Mall, London

One of the finest town houses in central London is that of Lord Schomberg, which was built in 1690 fronting onto Pall Mall and was embellished with Coade stone additions in 1790, comprising voussoirs, architraves and keystones (Fig. 7.54). The finest addition is the Coade stone portico displaying classical figures, shown here in Fig. 7.55, which is still much admired today.

Fig. 7.54 Lord Schomberg's House in Pall Mall, London, built in 1690 and applied with Coade stone embellishments *ca.* 1790

Fig. 7.55 Lord Schomberg's House, Pall Mall, London, the impressive classical portico in Coade stone, *ca.* 1790

7.19 The King George III Statue, Weymouth

The seafront Coade stone statue erected by the inhabitants of Weymouth, Dorset, to commemorate the 50 years of King George III's accession to the throne in 1809 is noteworthy for its survival in the otherwise corrosive marine atmosphere (Fig. 7.56). Designed by J. Hamilton, the inscription bears his name, an unusual addition for most Coade sculptures. It shows the King in his Garter robes standing in front of a table bearing his crown and a shield of arms standing on a tall plinth flanked by a life-sized lion and a unicorn. It serves to illustrate too that some Coade stone

Fig. 7.56 Weymouth promenade, Dorset, statue to celebrate the 50 years of King George III on the throne. (*Courtesy of Chris Talbot, CC by SA-2.0, Public domain*)

sculptures, like their natural stone counterparts were painted, sometimes to simulate a bronze effect: previously, we have mentioned the case of the Twickenham lion, which differs in that the application of gold paint to the lion statue occurred in the twentieth century (see Fig. 7.12). Its companion lion statue, the South Bank lion, although supplied in natural stone at first for Goding's Lambeth Brewery, was also painted red in the twentieth century for its relocation to Waterloo British Rail station concourse, but its later relocation to its current siting on Westminster Bridge involved the removal of the paint and restoration of its natural stone finish.

7.20 Twining's Tearooms, the Strand, London

A refurbishment of the Twining's Tearooms in The Strand, London revealed some Coade stone sculpture hidden under layers of grime and detritus (Fig. 7.57): the sculpture, dating from *ca.* 1790, comprises the figure of an Indian on the left and a Chinaman on the right, representing the two main types of tea that were available and imported by the East India Company ships in the late eighteenth century. Under the broken Adam-style pediment sits a Coade lion painted gold atop the Royal coat-of-arms of King George III. Thomas Twining opened his eponymous tea and coffee

Fig. 7.57 Twining's Tearooms, The Strand, London, showing the Coade stone figures of an Indian and a Chinaman representing the two main types of tea, with a Coade stone lion and the Royal warrant

shop here in 1706, at that time one of some 2000 coffee shops in London but he was distinctive in offering tea as an alternative beverage, despite the high taxation that was levied on tea at that time. Ladies were not admitted to the coffee houses, so Twining's tea house soon became a meeting place for ladies to take tea socially in London. As in the previous Case Study, the figures were painted, which may have assisted in their preservation in the Victorian heavily polluted London atmosphere.

7.21 Lady Gandes

In Fig. 7.58 is shown a neo-classical figure in Coade stone which is known as "Lady Gandes" and dated to 1794, height 1.75 m, width 0.6 m and depth 0.7 m. This is known as an "allegorical figure" as it does not match with other Coade stone replicas of ancient classical subjects from Greek, Roman or Egyptian literature or a mythical beast. It certainly does not represent the figure of a real Lady Gandes of the Georgian

Fig. 7.58 *Lady Gandes*, 1794, an allegorical Coade stone figure. (*Museum of Fine Arts, Houston, USA. The Rienzi Collection. Gift of Mr and Mrs Harris Masterson. Public domain*)

period. Nevertheless, she is a very beautiful rendition of a neo-classical statue and must therefore be considered somewhat unique in the annals of Coade stone topics.

7.22 Nelson's Columns

Following the Battle of Trafalgar, Horatio Nelson had five columns erected in his memory by public subscription involving Coade stone artefacts, all pre-dating Nelson's Column in Trafalgar Square, London (which dates from 1843 and which has atop an 18 feet high statue of Nelson in sandstone), whereas only one analogous version commemorating a British army commander in the Peninsular War exists: namely, that of General Sir Rowland Hill, see the next Sect. 7.23. Perhaps the best known of the Nelson columns fashioned with statues in Coade stone is that situated in Notre Dame Street, Montreal, later renamed the Place Jacques-Cartier (Fig. 7.59). It comprised a Coade stone statue of Nelson, 8 feet high on a limestone column of height 50 feet with Coade stone plaques celebrating his naval victories on its plinth, and was erected in 1809 following a call for a public subscription for a monument

Fig. 7.59 Nelson's Column, Notre Dame Street (Place Jacques-Cartier), Montreal. (*CCA-SA 3.0. Public domain*)

just after the Battle of Trafalgar: it may at first seem surprising perhaps that a predominantly French-Canadian city should wish to celebrate the destruction of a French Emperor's fleet at Trafalgar (which was described then as a "triumph of virtue") but the French Revolution and the removal of the monarchy was not popular in Canada at that time.

Nelson's Pillar in Sackville Street (now O'Connell Street), Dublin, was the first column to be erected by public subscription commemorating the Battle of Trafalgar, a granite column surmounted by a Coade stone statue of Nelson was erected on the 21st October 1809, the fourth anniversary of the battle of Trafalgar, at a total cost of £6586. Designed by William Wilkins, it stood on a square pedestal engraved with Nelson's greatest battles. The column was hollow and incorporated a spiral staircase of 168 steps leading to a viewing platform beneath the statue. The upper half of the column and statue were destroyed by a bomb in 1966, which was ascribed to the IRA or possibly an alliant faction group with the Basque separatists, ETA, although this was never proven. Some Coade stone pieces of the statue are still preserved in various museums in the city, including the head of Nelson. An engraving of the Nelson Pillar in Dublin prior to its destruction is shown in Fig. 7.60.

The Glasgow Nelson column (Fig. 7.61) on Glasgow Green dating from 1806 is the oldest commemorative statue related to Nelson still extant and is the first example of a civic monument funded by public subscription celebrating Nelson's victories. It was designed by David Hamilton and is 144 feet tall. It does not have a statue of Nelson but is in the form of an obelisk some 144 feet high and also serves to commemorate James Watt, the inventor of the steam engine.

Finally, the Nelson column at Great Yarmouth, known locally as the Britannia Monument (Fig. 7.62), was the last to be erected in 1819 at a total cost of almost £7000. Unlike some of the others it does not have a figure of Nelson, but rather a group of Victory caryatids in Coade stone bearing bronze wreaths and supporting a Coade stone figure of Britannia modelled by Joseph Panzetta and designed by William Wilkins at a cost of £892. It has suffered somewhat in the sea air and in 1896 some figures were replaced by cementitious copies and still later in the 1970s by fibreglass versions, which is a most unusual occurrence for Coade stone (and is reminiscent of our previous discussion in the case study of the Huddersfield Lion). The statue of King George III on Weymouth promenade that was erected some eight years earlier has seemingly not suffered so badly, despite its marine location: a possible explanation for this discrepancy resides in William Croggon's work notes (see Chap. 4), where it was noted that one of the figures for the Britannia Monument showed signs of cracking from the kiln firing, but he regarded this blemish as not serious and he went ahead with the installation of the commission anyway. It is suggested that perhaps this provides a rather rare example of Coade stone sculptures which have been installed despite having an inferior construction arising from a potential misfiring in the kiln, which would result in an incomplete mineralisation at high temperature as highlighted in Chap. 5.

Fig. 7.60 Nelson's Pillar, Sackville Street (O'Connell Street), Dublin. Engraving form 1829 by H/ Fisher, Son & Co., G. Petrie and R, Winkles. (*CC by NC—SA 3.0. Public domain*)

Fig. 7.61 Glasgow Green Monument, Glasgow, obelisk. (*CCA-4.0. Public domain*)

Fig. 7.62 Britannia Monument, Great Yarmouth, Norfolk. (*Courtesy of Bob Cook, CC by SA 2.0*)

7.23 Lord Hill's Column, Shrewsbury

As mentioned above, the only column topped by a Coade stone figure raised by public subscription to commemorate the victories of an Army commander in the Peninsular War, the statue to General Sir Rowland Hill (1772–1842), Viscount Hill, stands atop a 144 feet tall column in Abbey Foregate, Shrewsbury (Fig. 7.59). It was erected on July 18th 1816 to celebrate the first anniversary of the Battle of Waterloo, where Hill commanded the British II Corps and had the honour of receiving the surrender of Napoleon's elite Imperial Guard. Prior to that he had served with distinction in the major battles in the Peninsula at La Corunna, Talavera, Busaco, Vimiero, Salamanca and Vitoria. He eventually became Commander-in-Chief of the British Army in 1828. The *Gentleman's Magazine* of 1817 first noted the design with its engraving. Designed by Edward Haycock and modelled by Thomas Harrison (who also designed the pedestal supporting the column), it comprises a very large Greek Doric column of Grinshill stone some 134 feet high standing on a plinth and surmounted by a Coade stone statue of Lord Hill some 17.5 feet tall, modelled by Joseph Panzetta, the tallest single statue ever made of Coade stone. The town and county of Salop paid Eleanor Coade £315 for the statue. There are four lions situated at the base of the plinth which are made in the local Grinshill sandstone, which upset William Croggon who believed that he should have received the commission for these to be manufactured also in Coade stone; instead they were modelled by a local sculptor, John Carlile of Shrewsbury, in the local Grinshill sandstone. The fixing of the statue on its exceptionally tall column was costed at an additional £66 1s 8d. After firing the statue in Lambeth, Eleanor Coade put the statue on public viewing in her Manufactory in 1816.

The Lord Hill column is hollow, taller than Nelson's column in Trafalgar Square, and has 172 steps spirally arranged inside and the total cost of the whole structure was £5972 13s 2d, equivalent to about £250,000 today (Fig. 7.63).

7.24 Wolterton Hall, Norfolk

Wolterton Hall, a Palladian mansion built by Thomas Ripley for Horatio Walpole, the brother of Sir Robert Walpole, Prime Minister in 1727 and completed in 1742. Wolterton became the country seat of the Earls of Orford thereafter until the mid -nineteenth century. A vase, which is believed to be potentially of Coade stone and shown in Figs. 7.64 and 7.65, provided a specimen from this location for analysis in Chap. 6. It can be seen that this vase does not have the surface detail of the Borghese and Medici Vases discussed earlier and seems to lack evidence for the satyr's head handles possessed by these: also, serious environmental deterioration, delamination and spallation has occurred, which would potentially cast considerable doubt upon its veracity as a Coade stone relic.

Fig. 7.63 Lord Hill's Column, Abbey Foregate, Shire Hall, Shrewsbury. (*CCA-SA 2.0 credit: Keith Havercroft*)

Fig. 7.64 Wolterton Hall, Norfolk, front elevation. (*Courtesy of Caroline Stanford*)

Fig. 7.65 The urn at Wolterton Hall. (*Courtesy of Caroline Stanford*)

7.25 Syon Park, Isleworth, Middlesex

The Lion Gateway at Syon Park was designed by Robert Adam in 1767–9 on the commission of the Duke of Northumberland, whose London Home is Syon House, built on the site of the original mediaeval Sion Abbey. The gateway (Fig. 7.66) comprises a classical arch surmounted by the Percy lion; Robert Adam had originally commissioned another London artificial stone manufacturer, George Davy, to provide items in cold cast stone for use in the construction of the gateway but these did not survive so he recoursed to their replacement with Coade stone from Eleanor Coade in 1770/1. It is understood that several further replacements in Coade stone of deteriorated pieces on the gateway were undertaken in the early 1800s (Caroline Stanford, personal communication). A sample of suspected Coade stone from the gate pier (Fig. 7.67) was provided for analysis here.

Fig. 7.66 The Gateway at Syon Park, Isleworth. (*Courtesy of Caroline Stanford*)

7.26 St. Mary Magdalene Church, Madingley, Cambridgeshire

The church of St Mary Magdalene at Madingley dates principally from the thirteenth century, with an early fourteenth century north aisle and west tower of *c*.1300; a major restoration was undertaken between 1872 and 1874 by the architect J. Morley of Cambridge, and the chancel was heavily restored by Ewan Christian (Bradley & Pevnser, *The Buildings of England: Cambridgeshire*, 2014). Nearby is a large sixteenth century hall, built by Sir John Hynde between 1543 and 1547, which was subsequently occupied by his descendants until the 1860s. Eventually, after several succeeding owners, it became the property of Cambridge University in 1948.

The Royal Arms at Madingley Church is set on the south wall of the nave, above the south door within a moulded arch (Fig. 7.68).

The display of Royal Arms became widespread in parish churches during the reign of King Henry VIII, following his break with the Roman church, and were used principally to remind both the clergy and the worshippers that Parliament had

Fig. 7.67 The gateway pier at Syon Park which has had replacement of deteriorated original artificial stone by Coade stone. (*Courtesy of Caroline Stanford*)

confirmed the King as Supreme Head of the Church of England. They remain in use to the current day. The Arms at Madingley display those of King George III and are signed: "COADE & SEALY LAMBETH 1802" at the lower right. The identity of who installed it, or who executed the surrounding moulded arch, is unknown (Fig. 7.69).

The form of the arms comprise a two-tiered base with the Coade and Sealy signature and above this sits the escutcheon, oval in shape, bearing the arms of King George III, within which sits an inescutcheon depicting the arms of the House of Hanover. The main escutcheon is supported by a crowned lion and a unicorn, and at the top of the shield is a crowned lion passant guardant (Fig. 7.70).

The arms have been plastered with a backing of lime plaster. No evidence of paint pigment remains, if indeed it ever was painted, but white distemper has been applied uniformly. However, the correct heraldic colours are depicted by incisions on the surface (hatchings) implying pigment colour was unnecessary: here, the gold tincture is represented by dots, blue by horizontal bars and red by vertical bars. The condition of the arms shows that some parts have been detached and are now missing, and the ironwork within the surrounding arch is corroding and causing cracking. There is also an iron cramp behind the top of the shield, evidently tying it

Fig. 7.68 The Coade stone Royal Arms in Madingley Church, Cambridgeshire. (*Courtesy of David Carrington*)

Fig. 7.69 The COADE & SEALY LAMBETH 1802 impressed mark on the Royal Arms. (*Courtesy of David Carrington*)

Fig. 7.70 Detail of the escutcheon and inescutcheon on the Royal Arms enclosed within the Order of the Garter band, "Honi Soit Qui Mal y Pense". (*Courtesy of David Carrington*)

back to the wall behind. At the time of writing it is hoped these problems will shortly be addressed by conservation repairs (Carrington and Blood 2024).

Other examples of Royal Arms made from Coade stone are known and may once have been produced widely, however few examples now remain such as these arms at Madingley and those at Clerkenwell and Spitalfields in London, Petworth, Sussex, and at Tong, Shropshire.

References

R. Adam, C.H. Tatham, Thomas Hope's house in Duchess Street: the interiors created by Hope to display works of art in his London House were some of the most influential of the regency age. Apollo, May Issue (2004)

S.J. Bernard Burke, *Burke's General Armory—The General Armory of England, Scotland, Ireland and Wales: Comprising a Registry of Armorial Bearings from the Earliest to the Present Time* (Harrison & Sons, London, 1884)

G.G. Bottari, *Il Museo Capitolini*, Volumes 1–3 (Paolo Cavaletti/de Rossi, Rome, 1741)

S. Bradley, N. Pevnser, *The Buildings of England: Cambridgeshire*, 3rd edn. (New Haven/London, Yale University Press, 2014)

C.J. Brooke, H.G.M. Edwards, F. Kosek, J. Jehlicka, C. Stanford, M.J. Champion, The analysis of two historic Fired Art Stonewares: Coade stone (18th–19th centuries) and a recent discovery

from Oxnead Hall, Norfolk, (16th–17th centuries). Archaeometry, 1–11 (2022). https://doi.org/10.1111/arcm12842
D. Carrington, E. Blood, *St. Mary Magdalene Church Madingley, Cambridgeshire (Diocese of Ely): Conservation Survey Report on the Royal Arms (c.1802)* (Skillingtons, Grantham, 2024)
J. Claudius Loudon, *Encyclopaedia of Cottage, Farm and Villa Architecture and Furniture: Containing Numerous Designs for Dwellings from the Cottage to the Villa Including Farms and Other Agricultural Buildings and the Appropriate Office and Garden Furniture* (Longman, Beer, Orme, Brown, Green & Longman, Holborn, London, with Supplement, 1842)
E. Coade, J. Sealy, *The Description of the Grand Model of Neptune Giving up the Body of Nelson* (T. Romney, Survey Printing Office, Lambeth, 1813)
H.M. Colvin, *A Biographical Dictionary of British Architects, 1600–1840* (J. Murray, London, 1978).; 47th Edition, Yale University Press, USA, 2008
J. Cowan, *The Coade Stone Statue of George III: Conservation Report on the Bust and Statue Fragments*, vol 2008 (Lincoln Castle and Lincolnshire County Council, 2008), pp. 1–23
J.E. Cussans, *The Grammar of Heraldry: Containing a Description of All the Principal Charges Used in Armory, the Signification of Heraldic Terms and the Rules to be Observed in Blazoning and Marshalling* (Longmans, Green & Co., London, 1866)
I. Darlington, Coade's Artificial Stone in St George's Chapel, in *Annual Report of The Society of Friends of St George and the Descendants of the Knights of the Garter*, (Oxley & Son, Windsor, 1955), pp. 13–18
J. Fairbairn, *Crests of the Families of Great Britain and Ireland and Their Kindred in Other Lands*, Volumes I and II (T.C. and E.C. Jack/New Orchard Publishing, Edinburgh and London/London, 1905) and 1986
A.J. Finsberg, *Turner's Watercolours at Farnley Hall* (The Studio Publishing, New York, 1912)
P.F. Foggini, *Il Museo Capitolini*, vol 4 (Antonio Fulgoni, Rome, 1750)
Folly Flaneuse, The Dunston Pillar. *Ramblings to and Ramblings around Follies, Garden and Landscape Ornaments* (February 21st, 2020): the follyflaneuse.com
F. Haskell, N. Penny, *Taste of the Antique: The Lure of Classical Sculpture, 1500–1900* (Yale University Press, New Haven/London, 1981), pp. 315–316
T. Hope, *Household Furniture and Interior Decoration* (Longman, Hurst, Rees and Orme, London, 1807)
L.E. Karran, B.J. Colston, Coade, Blashfield or Doulton? The in situ identification of ceramic garden statuary and ornaments from three eighteenth and nineteenth manufacturers. J. Cult. Herit. **18**, 290–298 (2016)
A. Kelly, *Mrs Coade's Stone* (Self-Publishing Association/The Georgian Group, Upton-upon-Severn, Worcestershire, 1990)
A Kelly, ", Coade Stone: Its Character and Conservation, in *Architectural Ceramics: A Joint Symposium of English Heritage and the UK Institute for Conservation*, ed. by J.-M. Teutonio, 22–25 September 1994 (James Ltd Publishing, Manchester, 1996), pp. 32–38
H. Roberts, W.H. Godfrey, Lambeth Waterworks and the Lion Brewery, in *Survey of London*, (Lambeth: South Bank, Vauxhall, London, 1951), pp. 51–54
L. Roper, *The Gardens of Anglesey Abbey, Cambridgeshire* (Faber & Faber, London, 1964)
S. Scott, *Artificial Stone: A Successful Substitute for Natural Stone* (Haddonstone Ltd., Northampton, 2016)
L. Skerritt, *Condition Report and Assessment of the Immediate and Long-Term Needs of the Coade Stone Statue of George III*. Unpublished (Lincolnshire County Council, 2006)
C. Stanford, *The Lakeside Trust, English Heritage*, personal communication
C.H. Tatham, *Etchings of Ancient Ornamental Architecture Drawn from the Originals in Rome and Other Parts of Italy During the Years 1794, 1795 and 1797* (Thomas Gardiner, London, 1803)
The Builder, Volume 26, Issue 1328 (Building Publishers Ltd., 1868)
R. Thorne, *The History of Parliament, 1790–1820*, 5 volumes (Secker & Warburg, London, 1986)
D. Watkin, *Thomas Hope and the Neo-Classical Idea* (Murray, London, 1968)

D. Watkin, P. Hewat-Jaboor, *Thomas Hope, Regency Designer and Patron* (Bard Center for Studies in the Decorative Arts, Yale University Press, Hartford, 2008)

C.M. Westmacott, *British Galleries of Paintings and Sculpture: Comprising a General Historical and Critical Catalogue with Separate Notices of Every Work of Fine Art in the Principal Collections* (Sherwood, Jones & Co., London, 1824)

G.J. Williams, *Weston Park: The House, the Families and The Influence* (Boydell & Brewer Ltd, Suffolk, 2022)

Chapter 8
Conclusions

Abstract A conclusions chapter which gives a resume of the text of this book summarises the findings and also gives some proposals for future work in this area of artificial stone and its placement in the landscape and house history of the Georgian and Victorian periods.

Keywords Eleanor Coade recipe · Coade stone · Deterioration · Terracotta · Lithodipyra · Britannia memorial · King George III bust Lincoln Castle · Croggon · Richard Holt

In this book we have set out to examine critically the history of Coade artificial stone and its relationship with earlier artificial stone congeners and also some later fired ceramic successors which have been sometimes been generically named "terracotta" and their cold-cured moulded and cast cementitious analogues. On this journey we have identified several misconceptions, myths and unsubstantiated legends associated historically with Eleanor Coade and her eponymous stone artefacts. The survival capabilities of Coade stone ceramic statuary and building ornament in the environment relative to the softer sedimentary rocks such as sandstone, limestone, alabaster and marble is without question its outstanding merit that has been claimed correctly from its time of production and maintained through the ages, some 250 years since it was first produced by Coade & Sealy in the early 1770s.

Nevertheless, there are reports that perhaps there was some inbuilt variability in the ability of a Coade stone artefact to resist hostile environments, such as acid rain, sea fret, burial in building debris and temperature changes: this could reflect potential changes in its material composition brought on by sourcing problems of the raw materials or perhaps by inadequate kiln temperature control during the firing process. Both of these were experienced by Eleanor Coade's contemporaries in the fine porcelain industry, where manufacturers admitted to a significant wastage in their products amounting sometimes to greater than 25% of their output. These kiln misfires resulted in the formation of waste pits of unsaleable fired ceramic material which were often disposed of as "grog" elsewhere, and were used either in their own manufactory as a supplement to the normal raw materials or perhaps were ground up and sold on to other ceramic users to bulk up their own incipient ceramic paste

H. G. M. Edwards, C. J. Brooke, *Coade Stone*, Cultural Heritage Science,
https://doi.org/10.1007/978-3-031-71432-0_8

production. Eleanor Coade certainly was aware of this situation as her first description of her own artificial high-fired ceramic artificial stone was "Lithodipyra", literally meaning "twice-fired stone", referring to her use of 10% each of pre-fired grog and soda-lime glass from local glass works in her Coade stone, as indicated in her recipe published by Charles Fowler in the 1850s some years after the closure of the factory. Our comprehensive analyses of genuine Coade stone artefacts indicate that Eleanor Coade did not vary her paste composition significantly during the 70 years of operation of her manufactory and it seems that, although this has never been mentioned specifically, and indeed she never refers to the waste from her mis-fired kilns, she probably used her own grog waste materials in-house to supplement her raw materials in the manufacture of her artificial stone paste. Likewise, Eleanor Coade does not state the source of her grog and glass additives, unlike other components such as river sand and ball clay. Limited analytical data also indicate that the sourcing of her raw materials was also reliable, unlike other ceramics manufacturers who experienced problems which had a serious result upon their ability to produce their wares and in some cases caused the premature closure of their manufactories, such as the Limehouse Pottery in London which closed in 1748 due to its inability to source an alternative to the steatite obtained from Kynance Cove in Cornwall which ran out in 1747.

It has always been maintained that Coade stone artefacts were the epitome of resistance to environmental deterioration: even the lichen and cyanobacterial invasive biodeterioration observed as a "patina" on externally located Coade stone statues was not seen to be a problem for surface cleaning in this respect, unlike their marble analogues which were subject to a destructive hyphal penetration into the subsurface body of the stone occasioned by the complexation chemically of calcium which was sought by the lichen carboxylic acids in their metabolism for waste disposal. However, there are several instances where Coade stone relics exposed to hostile environments have been affected detrimentally and have need to be replaced not long after their original installation: an example of this is the Britannia memorial, where the four figures atop the tower exhibited destructive deterioration in the sea air and William Croggon needed to replace them forthwith. A possible explanation of this seemingly rare occurrence is that the particular firing sequence employed for the production of these figures was at fault—possibly a sufficiently high enough temperature could not be maintained accurately enough for the required time elapse to ensure the effective fusion of the raw materials .... this would certainly create internal faults and an imperfect substrate which would most likely be the centres for mineral damage environmentally. The specialist restorers and conservators of Coade stone relics now realise this and have commented that a range of qualities of artefact produced was potentially responsible .... the positioning of the paste artefact in the kiln, at the top or near the brick-lined furnace sides where it is hottest prior to firing was critical and many porcelain manufacturers were aware of the critical placement of their moulded shapes, which could be subject to firing cracks or alternatively "sagging" as a result of an imperfect temperature control. In the days prior to the use of accurate thermometry the skill of the kiln manager was vital in the control of the ceramics firing process.

The demise of Coade stone in the 1830s was quite rapid and has never been satisfactorily explained on one particular factor, although attributions have variously been made to the advent of cold-cured cast cementitious alternatives from 1828 which did not require firing and were hence much cheaper to produce: another possible explanation lies in the demand for *scagliola*, which was a specially formulated material cast onto wooden templates and dyed to simulate Italian marbles of several types and colours; this was a most desirable acquisition from the second decade of the nineteenth century and onwards. Certainly, William Croggon did diversify his Coade & Co. production from exclusively Coade stone artefacts into *scagliola* in the mid-1820s to supply the burgeoning industry and demand for this material for floors, columns, wall panels, table tops and furniture inserts: however, he was affected by serious business cash-flow problems due to bad debts and he was forced to declare bankruptcy in 1834, and soon dying in 1835. His son Thomas Croggon could not revive the business and eventually it was sold off in 1838 and the hardware, such as moulds, was dispersed in 1843.

In this text, we have sourced examples of genuine dated Coade stone and others which have an established provenance in house records and commission documents and form these we have evaluated the analytical composition of genuine Coade stone. It has also been discovered that Eleanor Coade and her successors did not change the formulation of her successful recipe through the production lifetime of the manufactory. The first formulation of Coade stone seems to match that of Daniel Pincot, as suspected, which dispels the idea proposed by some that Eleanor Coade must have made critical changes to Pincot's formulation upon taking over the artificial stone business in 1769. Equally clear is the revelation that Daniel Pincot's recipe and formulation was essentially different to that patented by Richard Holt and Thomas Ripley in 1722. Analysis of the interior of Coade stone artefacts also shows that Eleanor Coade did not adulterate her casting by using an inferior quality material upon which the top quality material was later applied—the composition of the artefacts is identical throughout.

Analytically, it has been shown that Eleanor Coade used a soda lime glass component in her raw materials and avoided the incorporation of lead oxide, which often was included in the "end-of-day" glass waste product from glass manufactories. This suggest that she did make a conscious effort to control the quality of her raw materials—as evidenced also by the low content of iron oxide present as impurities in her clays, flints and sand. It has long been recognised that the copying or replication of Coade stone artefacts in the mid- to late- nineteenth century was undertaken using cold-cured cementitious formulations, but these can rapidly be detected analytically because of their high content of calcia, which is at least an order of magnitude greater than is present in genuine Coade stone. Similarly, the use of gypsum plaster in repairs to Coade stone sculptures is noted by the presence of sulfur from the calcium sulfate deposits therein. Modern restorative procedures used in the conservation or reassembly of fragmented Coade stone sculptures are also detectable from their compositional variations and presence of resin adhesives. The most serious effect of modern restorations analytically is seen to be the masking of the potential Coade stone elemental oxide signatures and this was evidenced in the several specimens of the King George III restored bust and other fragments stored in Lincoln Castle: the insecure storage of the fragments over time has also contributed to a contamination problem that has been detected analytically.

The requirement of analytical sampling procedures involving the excision of specimens from artefacts is essential for the realistic assessment of elemental oxide compositions of artificial stone although it is appreciated that this is often not compatible with the wishes of conservators and owners of the artefacts concerned: the alternative route of using HHXRF instrumentation to non-destructively evaluate this information is appealing but has distinct reservation in its capability for the quantitative estimation of some lighter elements that paradoxically afford the key signals for confirmation of Coade stone. However, as for several other analytical applications in archaeology and museum science, the non-destructive *in situ* investigations can be treated as effective preliminary screening episodes for the eventual escalation into the strategic sampling by micro-excision of specimens of key artefacts. Apart from this, it is possible still to identify the cold-cured cementitious sculptures from their significantly larger signals of calcium oxide compared with Coade stone and especially when associated with detectable sulfur signals arising from gypsum plasters and mortars.

Appendices

Appendix I: The Genealogy of Eleanor Coade and the Coade Family History

In previous studies undertaken on armorial porcelains by one of the present authors (Edwards, *Welsh Armorial Porcelains: Nantgarw and Swansea Crested China*, 2022; Denyer, Denyer and Edwards, *The Pendock-Barry Service: A Forensic Evaluation*, 2023; Denyer, Denyer and Edwards, *Armorial Porcelain: The Genesis*, 2024) it was found that detailed genealogical research into the family backgrounds of clients, proprietors and decorators associated with the commissioning and manufacture of porcelains and ceramic earthenwares in the eighteenth and nineteenth centuries was invaluable for establishing the correct chronology of their manufacture and gave further insight into the potential reasons for the commissioning of the pieces concerned. The inclusion of genealogical data as part of the historical documentation component of a holistic forensic evaluation of a ceramic artefact along with scientific analytical studies and a stylistic connoisseurship was material to the establishment of the correct identification of important early nineteenth century services which hitherto had resisted proper definition on account of competing opinions based upon single unrelated pieces of information which had led to their misassignment. Genealogical data in their widest role can be accessed from newspapers, society columns and magazine announcements, wills, church monuments and memorials, gravestones, heraldic devices, crests and coats-of-arms, ancestor websites, local history of great houses and estate archives and legal proceedings: it is often possible to disprove conclusively apparently credible hypotheses relating to ceramic artefacts when genealogical data are considered as the factual chronology is not compatible with some expressed opinion.

Here, the broader Coade family history is compounded with the specific details of Eleanor Coade's genealogy to provide a definitive account of the personalities and people who were involved in her close circle when she embarked upon the synthesis of her eponymous artificial stone in the mid- to late- eighteenth century and onwards

H. G. M. Edwards, C. J. Brooke, *Coade Stone*, Cultural Heritage Science,
https://doi.org/10.1007/978-3-031-71432-0

into the early nineteenth century. The assimilation of this genealogical detail has been input into the true picture of events that occurred and has assisted in the dispelling of some incorrect assumptions and erroneous opinion which have hitherto been treated as factual historically.

I.1. The Coade Family History

Originally from Cornwall, the Coade family moved to Lyme Regis in Dorset in the seventeenth century. George Coade, Eleanor's great-grandfather, married Elizabeth Fowler and had a large family of 14 children: the eldest son, Robert Fowler Coade, married Sarah Enchmarch of Tiverton, Devon, and lived there prosperously in Great House, Broad Street, Tiverton. Their home was let during the summer to Lord Camelford, who owned mineral deposits in Cornwall that later provided the major sources of china clay and china stone for the burgeoning English ceramics industry in the eighteenth century. The discovery of kaolinite in Cornwall by William Cookworthy in the 1740s was a major landmark in the growth of the English porcelain industry and was material in his creation of the first English hard paste porcelain factory at Plymouth and later at Bristol.

The second son of George and Elizabeth, George, was born in 1706 and married Eleanor Enchmarch (born 1708) in 1732 and they then went to live in Exeter. Eleanor Coade was born there on the 24th June 1733 (her name is spelled Elinor in the records) and her sister Elizabeth was born in 1735. George Coade was a prosperous wool trader, specialising in cloth finishing, and contrary to some historians who place him living in Dorset, he lived in Exeter at 49 Magdalen Street for thirty years: he also bought a warehouse nearby to conduct his business locally. His brother Samuel became a partner in the business and he purchased Bunter's Castle in Lyme Regis, Dorset, as his family home, which he renamed Belmont House (Fig. AI.1), and which he left to his niece Eleanor Coade when he died in 1784. Belmont House has always been a prime site for the verification of Coade stone specimens as after its acquisition Eleanor used many of her artefacts to decorate it (Hansford, *Eleanor Coade of Lyme Regis*, 1960/61). Eleanor used Belmont House as her summer holiday villa and it is relevant that the first chemical analysis of Coade stone was accomplished using an authenticated specimen from Belmont House in 1991 (Freestone 1991). The latest analyses of Coade stone also used authenticated specimens from the gatepost pier in Belmont House (Brooke et al. 2022) The sister of Eleanor's mother, Mary Enchmarch, married a James Sealy, whose son, John, later became Eleanor Coade's partner in the new artificial stone firm of Coade & Sealy in 1772. Another aunt of Eleanor married Walter Oke Croggon of Grampound, Cornwall, who, following the death of John Sealy in 1813, then became an associate but not officially a partner of Eleanor Coade and later he took over as her successor in the firm of Coade & Sealy following Eleanor's death in 1821 when he purchased the business. William Croggon then carried on with the Coade stone business, which he purchased from Eleanor's estate in 1821 for £4000, as she had bequeathed her

Fig. AI.1 Belmont House in Lyme Regis, purchased by Samuel Coade, an uncle of Eleanor, who bequeathed it to her when he died in 1784. Eleanor used it as her holiday villa and decorated it profusely with Coade stone artefacts, including urns, voussoirs and keystones. Sometime later became the home of the author John Fowles. Now in the possession of the National Trust

whole estate to charitable causes, especially for the provision of succour and refuge for destitute women. Clearly, the family had a long connection with Devon, Dorset and the West Country and her cousin, John Coode (or Cood, an alternative spelling), was Mayor of Lyme Regis: he adopted illegally a coat-of-arms of argent, a chevron between three coots sable and acquired the crest of a coot sable (Roberts, *History of Lyme Regis and Charmouth*, 1834)—the coot crest appears on a Coade stone plaque on the gatepost pier (Fig. AI.2) at the entrance to Eleanor Coade's property, Belmont House, in Lyme Regis, and is believed to be a pun on the name, Coade = Coode = Cood = Coot. There are armigerous Coodes listed in Burke (Burke, *Burke's General Armory—The General Armory of England, Scotland, Ireland and Wales: Comprising a Registry of Armorial Bearings from the Earliest to the Present Time*, 1884) and Fairbairn (Fairbairn, *Crests of the Families of Great Britain and Ireland and Their Kindred in Other Lands,* 1905) but their crest is invariably represented as a Cornish chough sable and these are totally different from that of Eleanor Coade and her family.

In 1759, Eleanor's father, George Coade, became bankrupt: the wool trade was seriously in decline at this time and he was left with major debts. He moved with his wife, Eleanor, and daughters Eleanor and Elizabeth to Charterhouse Square,

Fig. AI.2 The gateway pier to Belmont House, Lyme Regis, showing the classical Coade stone paterae and the Coade stone crest plaque of a coot from the assimilated heraldic coat-of-arms

London, and then to Charles Square, Hoxton. In 1762 he was elected a Fellow of the Society for the Encouragement of the Arts and Sciences, now the Royal Society of Arts. Eleanor Jr. and her sister Elizabeth moved into 21, St Thomas Apostle's, and set up as linen drapers and cloth merchants in 1766; it has been suggested that the reason for this move out of the family home was so that they could keep their own household effects and finances distinct from those of their bankrupt father. Elizabeth's role seems to be that of keeping house for Eleanor and she died a spinster aged 40 in 1775. In 1769, George Coade became bankrupt again, having presumably discharged his debts earlier, and his property in Exeter was then sold: he died later that year in 1769. This is a mysterious event as it seems reasonable to conclude that George Coade had discharged his obligations from his first bankruptcy in 1759 but had seemingly engaged in some further unnamed business activities which had also been unsuccessful and led to his second bankruptcy—this suggests that Eleanor, an astute businesswoman, had potentially seen this coming and that this had encouraged her to separate her business activities and personal effects from those of her father earlier in 1766.

In 1769, Eleanor and perhaps also her mother, but not her sister, Elizabeth, purchased the ailing artificial stone business of Daniel Pincot in Lambeth and established the basis there of her Coade factory for the manufacture of artificial stone. Pincot in 1767 had set up a manufactory of artificial stone in Goulston Square, Whitechapel, offering similar artefacts to those which had been made by Richard Holt some thirty years earlier but "*executed in a Manner far exceeding anything of the like kind that has hitherto been offered to the Publick*". In October 1767, George Davy was also advertising artificial stone artefacts from his Goulston Square manufactory but Pincot had by then moved across the river to Narrow Wall, King's Arms Stairs, Lambeth: Davy was offering artefacts made in artificial stone cheaper than those which could be made in wood. Other small artificial stone manufactories were also active in London at this time, including one owned by a Mr. Bridges in Knightsbridge, which was also purchased by Eleanor Coade and later sold off in 1778. Other manufactories mentioned around this time which failed quickly were those of William Brooks, John Lynch, Samuel Arnold and Thomas Shape (Valpy and Kelly 1986). Pincot did not sell to his clients directly, although he did advertise a warehouse, and he offered his wares for sale through auctions in Pall Mall; there is a record that Lord Bessborough of Manorsea House purchased a quantity of Pincot's artificial stone through a Christies auction in 1767; unfortunately, for Daniel Pincot and the other manufacturers, these auctions did not yield economically viable prices for their artificial stone products, which were sold off at drastically cut prices. Soon after Eleanor's purchase of his business, Daniel Pincot was dismissed from his employment with Eleanor Coade in 1770 for alleged dishonesty in claiming publicly that he was the controlling partner and motivator behind her business. He died in 1792. Eleanor Sr, Eleanor Coade's mother, died in 1796.

Although the Coade manufactory was often referred to as Mrs. Coade's Artificial Stone Manufactory, especially in client invoices, one should not assume from this that Eleanor Coade Sr was involved in any way with the business: it was deemed polite in Georgian times to refer to a single businesswoman by the courtesy title of "Mrs". Eleanor Coade Sr appears in some documents as "Mrs Coade, Widow", which is unambiguous—and she certainly was the Mrs. Coade who appeared in the bills for the Poor Rate accounts for the property in which she lived with her daughters and for the Lambeth site as these were paid by the "Widow Coade". One document in particular sets the record straight: an invoice from a Miss Coade in February 1772 for £26-15s-6d sent to Burton-upon-Trent Town Hall for ornamental paterae to be supplied in her Coade stone. Another similar invoice was sent from Miss Coade to Hatfield Priory in 1771. These indicate that Eleanor Coade Jr. was the true proprietor of the Coade stone business.

Eleanor took on her cousin, John Sealy, as a business partner in 1771 and they later impressed their stone ware artefacts with a "COADE & SEALY, LAMBETH" mark which was sometimes accompanied with a date of manufacture; a variant on this mark was COADE LAMBETH, 1791, as borne by an Ionic column at Hill Hall (see Chap. 1 on the analysis of Coade stone marks) and COADE LAMBETH 1793 on a female statue at West Dean College, Sussex. John Sealy died in 1813 and the following year Eleanor took on William Croggon, another distant cousin, as a business associate but not as a partner, but she retained the Coade & Sealy appellation of her business for a while. As mentioned above, John Sealy was the son of her aunt, Mary Enchmarch; another sister had married Walter Oke Croggon and they had a son, William, who then became Eleanor's associate after John Sealy had died. In 1821 Eleanor died and William Croggon then bought the Coade & Sealy business from her estate, which had been left to charities, and ran it successfully until he encountered fierce competition from the newly emergent and much cheaper artificial stone variant based upon cold-cure Portland cement, which was not a high-fired ceramic and therefore did not require the expensive high-temperature curing stage in its production. He had diversified the business somewhat in the meantime and introduced the production of *scagliola,* a novel type of composite which simulated Italian and other pigmented marbles and was based upon a gypsum plaster compound with additives set and shaped on a wooden former. Lighter and with a more definite range of colours available than Italian marbles, *scagliola* was on the wish-list of many wealthy patrons who had completed the Grand Tour and who wanted neo-classical coloured marbles for their own great houses without incurring the heavy transportation and carving costs of the Italian marbles from Italy to England. Unfortunately, William Croggon had several bad debtors which resulted in his filing for bankruptcy in 1833 shortly before he died in 1835. His major debtor was the Duke of York, an avid enthusiast of *scagliola,* whose debts alone to Croggon were understood to be in the region then of £20,000 (estimated at several million pounds today). Thomas Croggon, William's son, took over the artificial stone business in bankruptcy for a while and it seems that artificial stone was still being supplied to commissions and orders up to 1837. The artificial stone business was not revived when Routledge and Greenwood (Routledge & Co.) made earthenwares on the site in 1837/8 and the artificial Coade stone moulds were sold off in 1843, many being purchased by Blashfield & Co. and by Mark Blanchard, who had both entered into the business of making artificial stone composites and who had been carefully monitoring the success of the Coade stone enterprise (Blashfield, *An Account of the History and Manufacture of Ancient and Modern Terracotta and its Use in Architecture as a Durable and Elegant Material for Decoration,* 1855). Actually, although most historians have ignored the reference, there is a record that during the last year of her life (in 1820–1821), Eleanor Coade and William Croggon did embark together upon the production of *scagliola,* mostly comprising architectural columns, but some were ordered by furniture dealers as was evidenced by the types of *scagliola* produced, namely narrow pieces, with front and side edges polished as would have been expected for their adoption into console tables and sideboards. London purchasers of Coade and Croggon's *scagliola* were the furniture makers Seddon of Aldersgate St., Ferguson of Oxford St., and John Mullane of Palace Row, New Road. Private purchasers were General Phipps, Earl Grosvenor, Sir John Langham and a Mr. Blacquire. Eight clock bases in *scagliola* were specially made by Coade & Co. for Messrs. Vulliamy of Pall Mall (Beard & Gilbert, *Dictionary of English Furniture Makers*, 1986).

As for their analogues of patterns in porcelain, Coade's artificial stone artefacts were reproduced and repeated over several decades of operation of the company: for example, the Coade stone Borghese Vase at Stourhead dated 1772 is exactly the same design as the Borghese Vase at Coleorton Hall, dated 1827. In addition, it was possible to add different motifs, or coats-of-arms and heraldic bearings, to standard artefacts at the behest of clients just as happened for their contemporary porcelains (Edwards, *Welsh Armorial Porcelains: Nantgarw and Swansea Crested China,* 2022). The Coade stone paste was particularly advantageous in this respect as intricate heraldic details could be incised with a stylus into the soft paste after moulding and before the kiln firing made them permanent. An example of this will be seen later in the Harlaxton Manor lions (ex- Witley Court) which hold the family escutcheons, and also the two falcon gatehouse pier crests of Walter Fawkes of Farnley Hall which bear his heraldic mascle symbol on their breasts. There was a continuous demand for neo-classical figures and even when there was a decline for Robert Adam's style of ornament and paterae for buildings that occurred with societal changes around 1800, this was replaced by an upsurge in demand for the classical garden statuary, ornaments and feature fountains. Eleanor Coade's stone received the commitment and support of most of the key architects of the day, some 148 in all as listed by Alison Kelly (Kelly, *Mrs Coade's Stone*, 1990), such as Sir John Soane (1753–1837), who following his Grand Tour in 1778, developed a lifelong enthusiasm for classical artefacts and he patronised the products of firstly Eleanor Coade and then William Croggon thereafter for 48 years continuously in his profession from the age of 27 in 1782 until 1830. Although several historians have alleged that the recipe and formulation for Coade stone was a well-protected secret (Kessels 1955), it was never patented and shortly after the demise of the manufactory full details of the recipe were published by Charles Fowler, a cousin of Eleanor Coade, in 1850 (Fowler 1850).

The wider Coade family were all Dissenters, who worshipped as devout Nonconformists and have been variously described as Baptists and Unitarians: as a result, they were not buried in Anglican cemeteries but in the Burnhill Field Cemetery, Islington, which was reserved for those of the Nonconformist faith (Dissenters). Certainly, George, Eleanor Sr, Eleanor Jr. and Elizabeth were all buried there in the family plot and historians have remarked that none have an artificial stone monument, for which Coade stone was renowned historically. However, the family vault occupied first by her father would, of course, have preceded the installation of a Coade stone monument and perhaps they did not see the necessity of modifying this later. Also, during the blitz of London in WWII, the Burnhill Cemetery was bombed by the Luftwaffe and V-1 flying bombs, causing serious damage, and an anti-aircraft battery was also sited there, the reverberations from which caused some additional damage to the surviving monuments. Hence, although Eleanor appears in the records of burials at the Burnhill Field Cemetery, her grave site was actually located in the bombed area and is now unmarked. Despite being a family member, John Sealy, Eleanor's business partner, must have been an Anglican as he is buried in St Mary's-at-Lambeth, his grave being marked by a plinth and monument in Coade stone, inscribed COADE & SEALY, LAMBETH. Daniel Pincot too was buried in an unmarked grave in the Burnhill Field Cemetery after his death in 1791. Burnhill Field Cemetery was an ancient burial ground for Nonconformists, having been opened in 1660 it was closed in 1853, some 130,000 burials later: some famous Nonconformists are buried there, whose monuments still survive, including John Bunyan, William Blake, Daniel Defoe and Susan Wesley (known as the Mother of Methodism).

References

G. Beard, C. Gilbert, eds., *Dictionary of English Furniture Makers, 1660–1840:* Eleanor Coade and Her Successors, London (1769- *ca.* 1840) (W.S. Maney, Leeds 1986)

S.J. Bernard Burke, *Burke's General Armory—The General Armory of England, Scotland, Ireland and Wales: Comprising a Registry of Armorial Bearings from the Earliest to the Present Time* (Harrison & Sons, London, 1884)

J.M. Blashfield, *An Account of the History and Manufacture of Ancient and Modern Terracotta and its Use in Architecture as a Durable and Elegant Material for Decoration* (1855)

R. Denyer, M. Denyer, H.G.M. Edwards, *The Pendock-Barry Service: A Forensic Evaluation* (Springer-Nature, Dordrecht, 2023)

R. Denyer, M. Denyer, H.G.M. Edwards, *Armorial Porcelain: The Genesis* (Springer-Nature, Dordrecht, 2024)

H.G.M. Edwards, *Welsh Armorial Porcelain: The Crested China of Nantgarw and Swansea* (Springer-Nature, Dordrecht, 2022)

J. Fairbairn, *Crests of the Families of Great Britain and Ireland and Their Kindred in Other Lands*, Volumes I and II (T.C. and E.C. Jack/New Orchard Publishing, Edinburgh and London/London, 1905 and 1986)

C. Fowler, Some remarks on terracotta and artificial stone as connected with architecture. Civil Eng. Archit. J. **13**, 215–216 (1850)

I. Freestone, Forgotten but not lost: the secret of Coade stone. Proc. Geol. Assoc. **102**, 135–138 (1991)

F.H. Hansford, Eleanor Coade of Lyme Regis, in *The Dorset Year Book* (1960–1961), p. 27

S.H. Kessels, Coade Stone: A Lost Secret. *Country Life Annual* (1955)

G. Roberts, *The History and Antiquities of the Borough of Lyme Regis and Charmouth* (Samuel Bagster & Sons, London, 1834)

N. Valpy, A. Kelly, Advertisements for Artificial Stone. Trans. Engl. Ceram. Cir. 206–226 (1986)

Appendix II: Eleanor Coade's First Catalogue of Artificial Stone Artefacts

The Catalogue was entitled "*A Descriptive Catalogue Of Coade's Artificial Stone Manufactory, At King's Arms Stairs, Narrow-Wall, Lambeth: Opposite White-hall Stairs.* It is fully referenced in Chap. 1 and was published in her name in 1784 and could be obtained at the Manufactory and at Mr. J. Strahan's, Bookseller, No.67, The Strand, London, and was priced at one shilling. It contained a listing of 778 individual standard items that could be ordered by clients, most of these being designed by the sculptor, John Bacon RA (1740–1799). The Frontispiece of the Catalogue is shown in Fig. AII.1. The contents page of the Catalogue listing is reproduced in Fig. AII.2, describing the categories of Coade stone artefacts that could be ordered and the first page is shown in Fig. AII.3, where 25 items of statuary are listed with prices: there followed another 29 pages of items up to No.746 and a further one-page Appendix listing another 32 items, bringing the total to 748.

The full Catalogue also came with a separate section which contained 36 leaves of etched designs of statuary and ornamentation compiled by Eleanor Coade and John Bacon between 1777 and 1779 and several of these plates were dated 1777 and 1778.

Fig. AII.1 Frontispiece of 1784 Catalogue for Coade's Lithodipyra or Artificial Stone Manufactory at King's Stairs, Narrow Wall, Lambeth, from a design by John Bacon RA

Copies of the Catalogue are available for consultation now in several places including, the British Library, the National Art Library of the Victoria and Albert Museum, Sir John Soane's Museum, in London and The Metropolitan Museum of Art, New York and the Yale Center for British Art (Paul Mellon Collection), New Haven, Connecticut.

A later catalogue appeared in 1799 under the authorship of Coade and Sealy and this contains approximately 200 more items than the 1784 version: many of these were added to the earlier Catalogue in handwritten manuscript in black ink. The provision of coats-of-arms as paterae or medallions in Coade stone was a generic standard issue, but of course these would be specifically designed according to the requirements of the armigerous client. An exception was made for the Royal coat-of-arms (No. 429 in the Catalogue) which was costed at 15 gns and the Prince of Wales feathers crest, No. 432) costed at 2 gns (£2 2s 0d). It is interesting that the *Borghese* and *Medici Vases* appear consecutively in the Catalogue at Nos. 81 and 82, and that both are recorded as being of an identical size, namely 4′ 2″ (1.27 m) in height, whereas measurements of these Coade stone vases now indicate that the *Borghese Vase* was significantly larger than its *Medici Vase* analogue and this can be discerned in their comparative setting on the *piano nobile* at Kedleston Hall; the *Borghese*

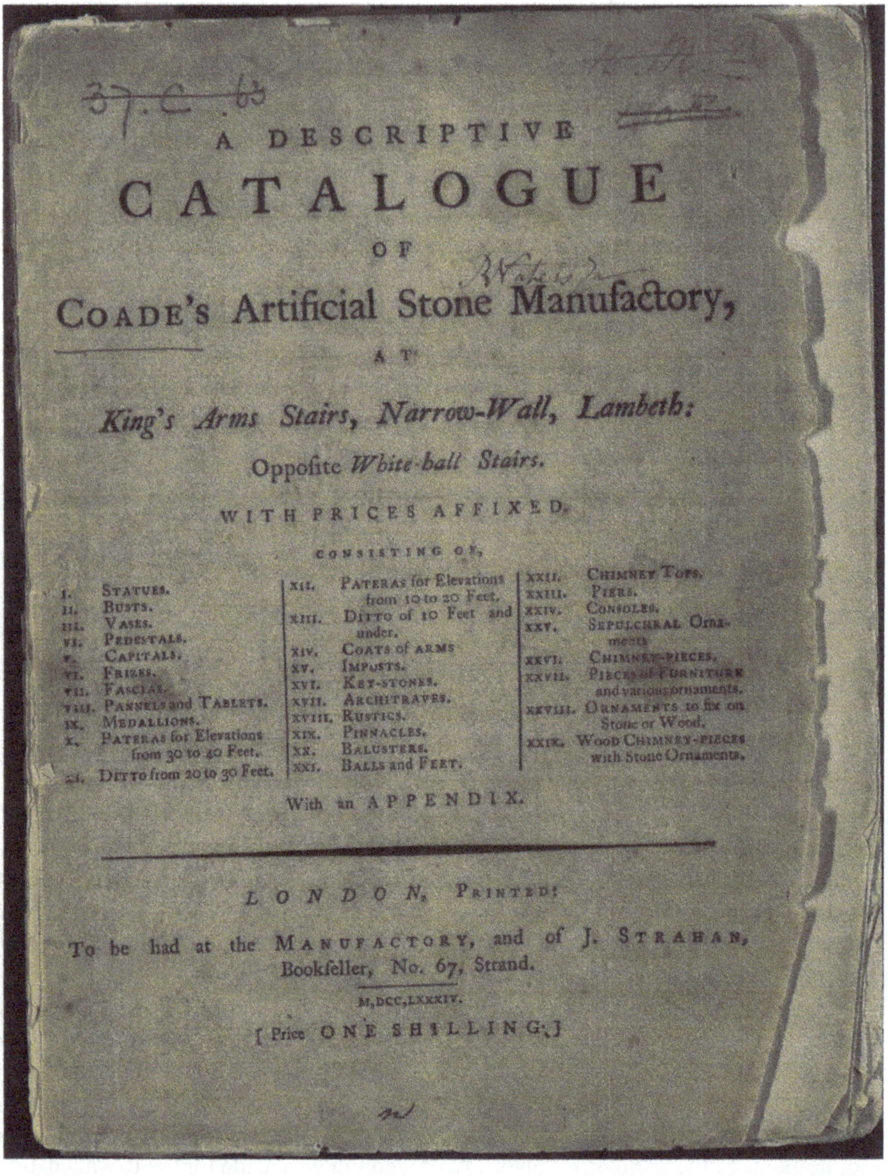

A DESCRIPTIVE
CATALOGUE
OF
COADE'S Artificial Stone Manufactory,
AT
King's Arms Stairs, Narrow-Wall, Lambeth:
Oppofite *White-hall Stairs.*
WITH PRICES AFFIXED,
CONSISTING OF,

I. STATUES.	XII. PATERAS for Elevations from 10 to 20 Feet.	XXII. CHIMNEY TOPS.
II. BUSTS.	XIII. DITTO of 10 Feet and under.	XXIII. PIERS.
III. VASES.	XIV. COATS of ARMS	XXIV. CONSOLES.
VI. PEDESTALS.	XV. IMPOSTS.	XXV. SEPULCHRAL Ornaments
V. CAPITALS.	XVI. KEY-STONES.	XXVI. CHIMNEY-PIECES.
VI. FRIZES.	XVII. ARCHITRAVES.	XXVII. PIECES of FURNITURE and various ornaments.
VII. FASCIAS.	XVIII. RUSTICS.	XXVIII. ORNAMENTS to fix on Stone or Wood.
VIII. PANNELS and TABLETS.	XIX. PINNACLES.	XXIX. WOOD CHIMNEY-PIECES with Stone Ornaments.
IX. MEDALLIONS.	XX. BALUSTERS.	
X. PATERAS for Elevations from 30 to 40 Feet.	XXI. BALLS and FEET.	
XI. DITTO from 20 to 30 Feet.		

With an APPENDIX.

LONDON, PRINTED:
To be had at the MANUFACTORY, and of J. STRAHAN, Bookfeller, No. 67, Strand.
M,DCC,LXXXIV.
[Price ONE SHILLING.]

Fig. AII.2 Contents page of the Descriptive Catalogue and Price List of 1784 for Coade stone artefacts: comprising 29 sections (see Chap. 2 for details) and an Appendix and listing 778 items

Vase is 1.72 m tall whereas the *Medici Vase* is 1.52 m tall, and both are therefore actually taller than their height as specified in the Catalogue. Of the 62 figures listed in the Catalogue none are cited for King George III, so most of those created in Coade stone must have been occasioned later by his Golden Jubilee celebrations of 1809. The sundial of Weston Park appears at No. 533 in the Catalogue at a cost of

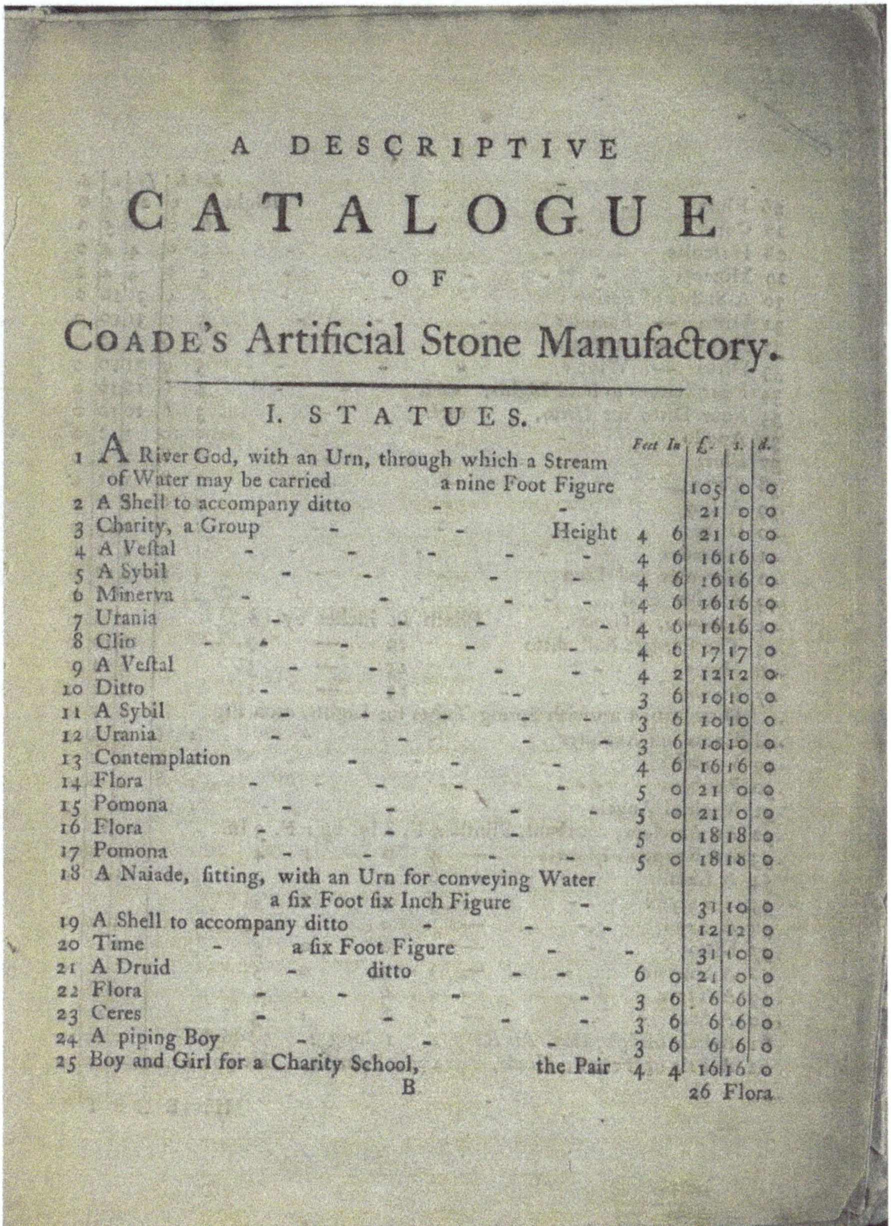

A DESCRIPTIVE

CATALOGUE

OF

COADE's Artificial Stone Manufactory.

I. STATUES.

		Feet	In	£.	s.	d.
1	A River God, with an Urn, through which a Stream of Water may be carried — a nine Foot Figure			105	0	0
2	A Shell to accompany ditto			21	0	0
3	Charity, a Group — Height	4	6	21	0	0
4	A Veſtal	4	6	16	16	0
5	A Sybil	4	6	16	16	0
6	Minerva	4	6	16	16	0
7	Urania	4	6	16	16	0
8	Clio	4	6	17	17	0
9	A Veſtal	4	2	12	12	0
10	Ditto	3	6	10	10	0
11	A Sybil	3	6	10	10	0
12	Urania	3	6	10	10	0
13	Contemplation	4	6	16	16	0
14	Flora	5	0	21	0	0
15	Pomona	5	0	21	0	0
16	Flora	5	0	18	18	0
17	Pomona	5	0	18	18	0
18	A Naiade, ſitting, with an Urn for conveying Water — a ſix Foot ſix Inch Figure			31	10	0
19	A Shell to accompany ditto			12	12	0
20	Time — a ſix Foot Figure			31	10	0
21	A Druid — ditto	6	0	21	0	0
22	Flora	3	6	6	6	0
23	Ceres	3	6	6	6	0
24	A piping Boy	3	6	6	6	0
25	Boy and Girl for a Charity School, the Pair	4	4	16	16	0

B 26 Flora

Fig. AII.3 First page of the Descriptive Catalogue and Price List of 1784 for Coade stone artefacts listing 25 statues, commencing with the River God at No.1

18 gns. and the pineapple at Ham House is No. 127, costing 2 gns. Perhaps the most surprising omission from the Catalogue of 1784 is the much admired sculpture of the *Dying Gladiator* or *Dying Galatian*, this ancient Greek statue from about 230 BCE

Fig. AII.4 Bronze statue of Dying Gaul, or Dying Galatian, at Iford Manor, Wiltshire, by Peter Scheemaker, *ca.* 1740. The original marble sculpture in the Capitoline Museum, Rome depicts a barbarian warrior in his death throes from the battle between the Greeks and the Celtic Gauls of Anatolia in 230 BCE and his dignity made a lasting impression upon visitors on the Grand Tour in the eighteenth century. (*Public domain, CCA-3.0*)

required a mandatory visit to the Capitoline Museum in Rome for participants of the Grand Tour in the eighteenth century (Fig. AII.4). This will be discussed later in a case study (Chap. 7) in connection with the two statues of the *Dying Gladiator* flanking the entrance staircase to Wentworth Woodhouse in Yorkshire.

The publication of the later Catalogue coincided with the opening of Coade & Sealy's Gallery, displaying examples of their fired ceramic wares, which could be entered at the corner of Pedlar's Acre and Narrow Wall at the southern end of Westminster Bridge.

Appendix III: Comprehensive Literature on Coade Artificial Stone and Related Materials

There now follows a list of articles and books which predominantly address the subject of Coade and other artificial stones: it has been found useful to collate these as part of the research undertaken in the writing of this text and a comprehensive list will be useful for readers who wish to study some aspects of artificial stone in a greater depth. The comment has been made by several authors and historians that the

Victorians did not appreciate that Coade's artificial stone artefacts were not natural stone, so realistic was their imitation, and that, therefore, after the closure of the Coade manufactory in the late 1830s the whole concept was forgotten until the early to mid-twentieth century when the myths and legends about Coade stone's success were proliferated. There is no doubt that Alison Kelly has made the greatest impact upon our understanding of Coade stone in particular through her comprehensive research, which over the period of twenty years collected together much of what we currently know about the individual artefacts and their original locations—thereby identifying many of the lost pieces in the process and several which had been relocated. Kelly relied, of course, upon many of the specific articles listed here and especially accounts of local historians and chroniclers of the great houses, which also generated several of the myths that surrounded this mysterious synthetic material and the persons associated with it. Local historians, as we have recounted here on several occasions, often only presumed that an artefact was made of Coade stone, possibly from local hearsay, and in some cases it has proved not to be so: the choice in Georgian times was quite clear—Coade stone or natural—but later historians did often confuse an artificial stone artefact with a natural stone analogue, whether ceramic or cementitious, as variants of the synthetic material from competitors and followers started to appear in their surveys. The initial distinction was necessarily entirely visual and mistakes could occur: the analytical support for a Coade stone assignment and its definitive attribution has only appeared, and then only rather sparsely, in the last three decades and analysed exemplars are still not numerous; it is it to be expected that some novel advances can be made in this direction in future projects which will assist in the production of a definitive database of genuine Coade stone specimens. Even when supported by house or estate documentation and invoices in their archives relating to the purchase, delivery and installation of Coade stone pieces, this evidence forensically does not automatically impose an authenticity upon the particular object under investigation as it has been found that as the Coade stone pieces replaced damaged stone artefacts, the replacement of original Coade stone items by later versions in natural stone and in cementitious stone are well documented too. Hence, consultation of as much relevant literature as available is mandatory in such a research project and the motto of the Royal Society of London from the mid-seventeenth century is very applicable in this context: *Nullus in Verbum — Believe No One*! This was a mantra devised by early scientific practitioners such as Robert Hooke and Sir Isaac Newton to dispel the indiscriminate ideas based upon the statement of misguided opinions and proposed as fact by their colleagues and to generate the truth independently using scientific methods. Isaac Newton believed that there could be no substitute to replace experimental data when advancing a hypothesis—"*Hypothesis non fingo*"—and this will serve as an admirable baseline for the determination and direction of future Coade stone analyses to establish a caucus of genuine exemplars. The next stage would be evaluate the many historical suggestions of statuary being assigned as Coade stone, often on an allegorical basis, some of which appear in the literature cited below.

References

G. Beard, C Gilbert, eds. *Dictionary of English Furniture Makers, 1660–1840:* Eleanor Coade and Her Successors, London (1769- *ca.* 1840) (W.S. Maney, Leeds, 1986)

M. Bimson, Stoneware in the 16th–20th centuries. Trans. Engl. Ceram. Cir. (1980)

J.M. Blashfield, *An Account of the History and Manufacture of Ancient and Modern Terracotta and its Use in Architecture as a Durable and Elegant Material for Decoration* (1855)

C. J. Brooke, H.G.M. Edwards, F. Košek, J.Jehlička, C. Stanford, M.J. Champion, Analysis of two historic fired artificial stones: Coade stone and a recent discovery from Oxnead Hall, Norfolk. Archaeometry, 1–11 (2022)

H. Clay, *Coade's artificial stoneware* (The Connoisseur, LXXXIII, 1928), pp. 79–87

E. Coade, J. Sealy, *A Descriptive Catalogue of Coade's Artificial Stone Manufactory at King's Arms Stairs, Narrow Wall, Lambeth: Opposite Whitehall Stairs. With Prices Affixed* (Manufactory and J. Strahan, London, 1784)

F.J. Collins, Coade stone. *Museum Journal***57/2**(May 1957), 37 (1957)

H.G.M. Edwards, *Porcelain to Silica Bricks: The Extreme Ceramics of William Weston Young, 1776–1847* (Springer -Nature, Dordrecht, 2019)

H.G.M. Edwards, *Porcelain Analysis and Its Role in the Forensic Attribution of Ceramic Specimens* (Springer -Nature, Dordrecht, 2021)

K. Esdaile, Coade Stone. Archit. Build. News **CLXI**(19th January 1940), 94–96 (1940a)

K. Esdaile, Coade Stone. Archit. Build. News **CLXI**(26th January Issue), 112–114 (1940b)

K.A. Esdaile, *English Church Monuments, 1510–1840* (B.T. Batsford Ltd., London, 1946)

Etchings of Coade's Artificial Stone Manufactory at King's Arms Stairs, Narrow Wall, Lambeth, 1784, British Library.

A. Fontaine, *Der Englisch Coade -Stein die Villeroy & Boch "Terrakotta": mite in Exkursen zu Martin Klauer nach Weimar and Friedrich Schinkel nach Berlin* (Books on Demand, BoD Alemana, Espana, 2019)

C. Fowler, Some remarks on terracotta and artificial stone as connected with architecture. Civil Eng. Archit. J. 13, 215–216 (1850)

A.J. Francis, *The Cement Industry, 1796–1914: A History* (David & Charles, Newton Abbott, 1977)

I. Freestone, Forgotten but not lost: the secret of Coade stone. Proc. Geol. Assoc. 102, 135–138 (1991)

I. Freestone, M.C. Bimson, M.S. Tite, The constitution of Coade stone, in *Ancient Technology to Modern Science, Ceramics and Civilisation*, ed. by W.D. Kingery, vol. 1, (Columbus, American Ceramics Society, 1985), pp. 293–304

I. Freestone, M.C. Bimson, M.S. Tite, Some Recent Research on Coade Stone. Trans. Engl. Ceram. Cir. **12**, 203–205 (1986)

R. Gunnis, *Dictionary of British Sculptors, 1660–1851*, 1953. Reprinted by Gale ECCO (Eighteenth Century Collection Online) (Detroit, Michigan, USA 2010)

S.B. Hamilton, Coade Stone. Archit. Rev. **CXVI**(November 1953), 295–299 (1953)

F.H. Hansford, Eleanor Coade of Lyme Regis, in *The Dorset Year Book* (1960–1961), p. 27

R. Holt, *A Short Treatise on Artificial Stone as 'tis Now Made and Converted into All Manner of Curious Embellishments and Proper Ornaments, of Architecture,* London, 1730; reprinted by Gale ECCO (Eighteenth Century Collection Online) (Detroit, Michigan, USA 2010)

M. Howard, E. Town, The development of Oxnead: from Tudor house to Nicholas Stone, in *The Paston Treasure: Microcosm of the Known World*, ed. by A. Moore, N. Flis, F. Vanke, (Yale University Press, New Haven, 2018), pp. 126–139

B.L. Hurst, Concrete and the structural use of cements in England before 1890. Proc. Inst. Civil Eng. Struct. Build. **116/3** (1996)

L.L. Jewitt, *The Ceramic Art of Great Britain from the Prehistoric Times Down to the Present Day*, vol I and II (Virtue & Co., London, 1878)

L.E. Karran, B.J. Colston, Coade, Blashfield or Doulton: the *in situ* identification of ceramic garden statuary and ornament from three 18th century and 19th century manufacturers. J. Cult. Herit. **18**, 290–298 (2016)
A. Kelly, Decorative Stoneware in Disguise. *Country Life*, 29th November 1973 Issue, 1797–1800 (1973)
A. Kelly, Mrs Coade's Gothic. *Country Life*, 2nd June 1977 Issue, 1514–1516 (1977)
A. Kelly, Mrs Coade's stone. *Connoisseur***CXCVII**(January 1978), 14–25 (1978)
A. Kelly, Mrs Coade's stone – stoneware, statues and ornaments. *Antique Collector*, August 1981 Issue (1981)
A. Kelly, An Expensive Present: The Adam Screen in Rio de Janeiro. Burlingt. Mag. **126**-(September Issue) (1984)
A. Kelly, Coade stone in Georgian Architecture. Archit. Hist. 71–101 (1985)
A. Kelly, Coade stone interiors. Antique Collect. July 1986 issue, 50–55 (1986)
A. Kelly, Coade Stone in Georgian Gardens. J. Gard. Hist. Soc. **16**, 109–133 (1988)
A. Kelly, Sir John Soane and Mrs Eleanor Coade. *Apollo*, April 1989 issue (1989)
A. Kelly, *Mrs Coade's Stone* (Self-Publishing Association, Worcestershire, 1990)
A. Kelly, Coade stone: its character and conservation, in *Architectural Ceramics*, (English Heritage, 1996)
A. Kelly, Eleanor Coade, in *Oxford Dictionary of Biography*, (Oxford University Press, Oxford, 2004). https://doi.org/10.1093/ref.odnb/37296
S.H. Kessels, "Coade Stone: A Lost Secret", Country Life Annual, 1955
R. Knowles, *What Regency Women Did for Us* (Pen & Sword Publishing, Barnsley, 2017)
D. Pincot, *An Essay on the Origin, Nature, Uses and Properties of Artificial Stone* (1770)
M. Ponting, *Investigation of Coade Stone Fragments from West Dean College, Sussex*, Historic England Research Report/Centre for Archaeology Report, 93/2003 (English Heritage, London, 2003)
G. Roberts, *The History and Antiquities of the Borough of Lyme Regis and Charmouth*, Samuel Bagster & Sons, London, 1834.
J.P. Ruch, Regency Coade; a study of the Coade record books, 1813–1821. Archit. Hist. **11** (1968)
S. Scott, *Artificial Stone: A Successful Substitute for Natural Stone?* (Haddonstone Ltd, Northampton, 2009)
C. Stanford, Revisiting the Origins of Coade Stone. Georgian Group J. **XXIV**, 95–116 (2016)
M. Stratton, *The Terracotta Revival* (Victor Gollancz, London, 1993)
S. Swann, *The Building Conservation Directory: 19th Century Cementitious Sculptures and Rockworks* (Cathedral Communications Ltd, 2019)
J. Tavenor-Parry, *An episode in the history of English Terracotta* (Archit. Rev., XXXIII, 1913), pp. 119–122
N. Valpy and A. Kelly, "Advertisements for artificial stone", Trans. Engl. Ceram. Circ. 206-226, 1986
H. van Lemmen, *Architectural Ceramics* (Shire Publications Ltd., Princes Risborough, 2001)
H. van Lemmen, *Coade Stone* (Shire Publications Ltd., Princes Risborough, 2006)

Appendix IV: The Lion of Huddersfield

Here we shall include some more information and archival photographs of the Lion of Huddersfield which forms the case study subject of Sect. 7.2 described in the main text.

The West Yorkshire Archaeology Advisory Service (MWY 7301) confirm in a report that the original nineteenth century sculpture atop the Lion Chambers and

Arcade in Huddersfield was replaced in 1977 by a fibreglass version: the pictures taken by the site team at that time in 1976 for assessment, just prior to its removal and demolition, illustrate the parlous state and extremely dangerous condition of the Huddersfield Lion as shown in the photographs in Figs. AIV.1, AIV.2, AIV.3, AIV.4, AIV.5, AIV.6, and AIV.7. These open with a view of St George's Square (Fig. AIV.1) and the lion atop the parapet and pediment of the Lion Buildings in 1976: a close inspection made of the Lion in the following photographs compared with the earlier ones shown in Chap. 7 Figs. 7.2 and 7.3 reveals that the sculpture has had to be shored up with scaffolding, the presence of severe cracks and deep faults can now be seen in the body of the lion, the excessive wear to the surface evidenced by a de-lamination, pitting and the loss of structural definition, and the requirement of four metal strengthening armature straps around the body that were needed to support the central framework can be seen (Figs. AIV.2, AIV.3, AIV.4, AIV.5, AIV.6, and AIV.7). Figure AIV.1 shows the Lion still atop its plinth but with metal armatures fitted and before the scaffolding was applied. It is hardly surprising that

Fig. AIV.1 The first of seven photographs from the archives of the West Yorkshire Joint Services, Archaeology Advisory Service, dated to *ca.* 1976, showing the location of the statue sitting atop its pediment on the Lion Buildings, St George's Square, Huddersfield. (*Reproduced with permission and courtesy of Rhona Finlayson, Historic Environment Records Officer, WYJS, Morley, Leeds*)

Fig. AIV.2 First of six photographs taken by an archaeological assessment team in *ca.* 1976 to verify the condition of the Lion statue on its plinth atop the pediment of the Lion Buildings: this side view shows clearly several large fissures that have appeared on the front and back legs and the provision of metal armatures encasing the body of the lion. (*Reproduced with permission and courtesy of Rhona Finlayson, Historic Environment Records Officer, WYJS, Morley, Leeds*)

Fig. AIV.3 Picture of the lion statue form the front, showing serious cracks across the face and damage possibly exacerbated by the first abortive attempt to lift it into position in February 1853. There is some evidence here and in the following photographs of delamination and pitting, causing loss of definition and detail. (*Reproduced with permission and courtesy of Rhona Finlayson, Historic Environment Records Officer, WYJS, Morley, Leeds*)

Fig. AIV.4 Picture of the lion's head from the right hand side, showing cracks across the leg and face with evidence of delamination. (*Reproduced with permission and courtesy of Rhona Finlayson, Historic Environment Records Officer, WYJS, Morley, Leeds*)

Fig. AIV.5 The lion's head from the left hand side, with multiple cracks across the legs and body and showing the armatures fitted to strengthen the body. (*Reproduced with permission and courtesy of Rhona Finlayson, Historic Environment Records Officer, WYJS, Morley, Leeds*)

the Lion of Huddersfield was not dismantled very carefully and saved in its entirety as its condition would probably have resulted in its fragmentation and collapse during removal from its parapet and plinth.

The photograph in the archive in Fig. AIV.1 in 1976 is similar to that shown in Fig. 7.3 in the main text in that it shows the deteriorated sculpture sitting upon its plinth, prior to the fitting of scaffolding, above the date of 1853, and with the iron armatures fitted as reinforcement around its body, but a high resolution magnification of this image (not shown) reveals that LEO is impressed into the plinth just below the right front paw of the Lion. This was obviously known locally for many

Fig. AIV.6 The lion statue from the left hand side rear, showing cracks on the front and rear leg and also on the body in front of the armature. (*Reproduced with permission and courtesy of Rhona Finlayson, Historic Environment Records Officer, WYJS, Morley, Leeds*)

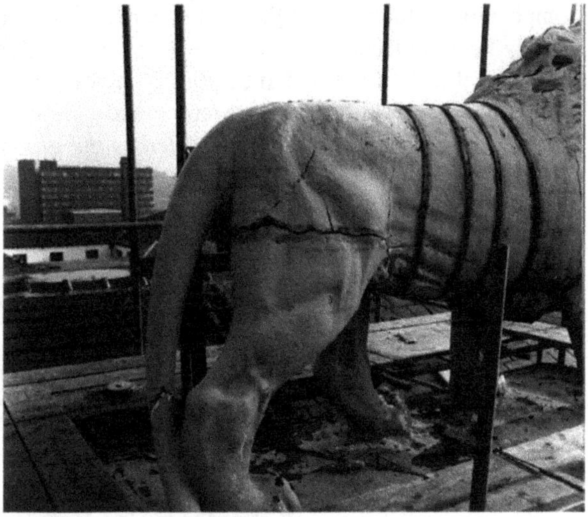

Fig. AIV.7 Rear of lion from the right hand side, showing large cracks and fissures to the body and legs, and evidence of potential atttempts made to seal these historically. The floor of the plinth upon which the lion stands also seems to be in a parlous state od deterioration. (*Reproduced with permission and courtesy of Rhona Finlayson, Historic Environment Records Officer, WYJS, Morley, Leeds*)

years as Professor Simon Armitage, Professor of Poetry at the University of Leeds (The Millennium Poet) and elected the 21st Poet Laureate of the United Kingdom in 2019, wrote a poem about "Leo", which he recalled from his childhood days in Huddersfield (Armitage, *Selected Poems*, page 129, 2001).

In common with the report of Historic England referred to in the main text, the original sculpture is still referred to as Coade stone but in the records in one instance (West Yorkshire Historic Monument Report HER MWY 7301), it has the additional appendage "*Coade stone (concrete)*" applied on the second page which is even more questionable, although this document does confirm that it was purchased from John Seeley of London in 1853.

At the present time, the Lion of Huddersfield still sits atop its plinth above the Lion Arcade and Lion Chambers although this is the fibreglass replacement, seemingly coated with a cementitious rendering, which nevertheless is a faithful copy of the original as can be seen in Fig. AIV.8. Figure AIV.9 shows the other lion on this structure which is a completely different heraldic representation: whereas the Lion of Huddersfield on the rooftop plinth could be described as a lion regardant, its counterpart above the street front door of the Lion Chambers three floors below is a lion couchant, displaying a more prolific mane, crossed front paws and assuming a different tail position!

Local history of Huddersfield has been comprehensively surveyed by Gordon and Enid Minter in a five volume text published between 1993 and 2002, which has now been digitally compiled by the Huddersfield Historical Society (G. and E. Minter, *Discovering Old Huddersfield*, 2010). In Volume I of this text the Lion Building in St George's Square has a brief mention and the squabbles between Samuel Oldfield the builder, James Pritchett the architect and Sir Willam Ramsden, a local landowner, are highlighted especially those that occurred between 1852 and the official opening of the Lion Buildings in January 1854. The Minters do not mention the composition of the original lion statue, which of course by their time of writing had already been replaced with the fibreglass version. The *Huddersfield Chronicle*, on the 10th February 1853, noted that the lifting of the lion onto its parapet was fraught with danger as the lifting chains broke and "*the enormous mass (authors: 7 tons according to supporting documents) fell to the ground when raised to a height of*

Fig. AIV.8 The Lion of Huddersfield as it is today: a fibreglass replacement version of the original, dating from 1977, but still sitting on its sandstone plinth above the Lion Arcade, dated 1853

Fig. AIV.9 The gilded lion couchant figure above the entrance door to the Lion Arcade and Lion Chambers, Huddersfield

seven feet causing damage to the nose and cheek of the royal brute", from which we can infer that the mechanical shock cannot have assisted in the eventual preservation of the statue's integrity. This was soon followed with a laudable description in the same newspaper of the second and successful attempt to elevate this statue onto its plinth and pediment, noting that it was "*A noble looking piece of work which will have a very imposing appearance when raised to its destined position*"!

From the West Yorkshire Kirklees Archives held at Huddersfield, documents cited as KC1/61–74 and KC1/84 apparently refer to the contracts, bills and accounts rendered for the Lion Buildings in Huddersfield, but in fact these all relate actually to the eponymous Lion Buildings in Northgate, Cleckheaton which were designed by Howorth & Howorth, architects and surveyors of Cleckheaton (personal communication, Rosie Hall, Archivist, West Yorkshire Archives, Huddersfield). It is not known whether the Lion Buildings in Cleckheaton also included a lion sculpture but clearly these documents are not relevant to our particular quest for information about the Lion Buildings in Huddersfield. Furthermore, a search carried out for a potential Cleckheaton Lion revealed that in 1785 a lion sculpted in local stone was mounted over the entrance archway to the estate at Pyenot Hall, Cleckheaton, which was known as the "Pride of Pyenot" (Fig. AIV.10). Pyenot Hall (Fig. AIV.11) was constructed as a family home in 1785 by Henry Birkby, a prosperous local card maker for the woollen industry, who also owned Rawfords Mill in Liversedge which was bankrupt by 1811 and thereafter leased by William Cartwright until it was attacked by Luddites in April 1812. Thereafter the Goldthorp family, also in card

Fig. AIV.10 The lion of Pyenot Hall standing atop the entrance archway to the estate. (*Public Domain*)

making, lived at Pyenot Hall until 1891, when Guy Goldthorp of Messrs. R. & G. Goldthorp disappeared, leaving creditors and the firm in bankruptcy. The Hall and estate were then bought by Enos Smith of Halifax, followed by Jonas Haley of Cleckheaton, and the card making business revived and extended into wire-drawing under the name Bridon Wire. After the collapse of Bridon Wire, the Hall became derelict and was destroyed by fire, being demolished in 1990 to make way for a new housing development along Bridon Way. The Spen Valley Civic Society searched for the missing lion of Pyenot Hall, which had been toppled from its arch and was later found in undergrowth, being restored to its place in the new housing estate (Helen Mead, *Telegraph and Argus*, 24th December, 2022), where it is now known locally as "Yon Lion"(Fig. AIV.12).

Fig. AIV.11 Pyenot Hall, Cleckheaton, undated. (*Public Domain*)

Fig. AIV.12 The stone Lion of Pyenot today, known as "Yon Lion", in Bridon Way, Cleckheaton. (*Public Domain*)

References

S. Armitage, *Selected Poems* (Faber & Faber, London, 2001), p. 129

G. and E. Minter, *Discovering Old Huddersfield*, 5 Volumes, Hollingworth & Moss, Leeds, 1993–2001, condensed and digitalised by the Huddersfield Historical Society in 2010.

R. Hall, West Yourkshire Kirklees Archives, Huddersfield, personal communication.

H. Mead, How a Stately Lion Came Back to Life in Suburbia. *Telegraph and Argus* (24th December, 2022)

Glossary

Biscuit porcelain and stonewares The unglazed product of the high-temperature firing process in the manufacture of porcelain and stonewares using a "biscuit kiln" which operated typically at a temperature in the region of 1100–1400 °C. The resultant ceramic is of a pure creamy-white or ivory colour and possessed a texture which, if blemish-free and perfectly shaped, was very highly prized by ceramic artists and modellers particularly for the construction of ornamental figurines in simulation of marble statuary. In the mid- to late -eighteenth century, the finest biscuit porcelain artefacts were commonly placed on dining tables for admiration by the guests and to stimulate conversation during dinner.

Blemish A blemish is a fault in the fired porcelain or stonewares, whether this is in biscuit or glazed form, arising from several problems which could arise during the firing process such as inadequate kiln temperature control leading to firing cracks, the introduction of contaminants such as carbonaceous matter which then burned to leave soot deposits and the emission of gases during the firing process which created bubbles in the vitrified paste or the glaze. Serious blemishes resulted in ceramic items being unsaleable.

Body paste The wet mixture of raw material components which comprise the formulaic recipes for ceramic bodies. Texturally moist to the touch, the paste could be moulded or formed into shapes preparatory to the firing process taking place in the biscuit kiln. This term is also applied to the fired ceramic body prior to its chemical analysis, when of course it is then not a wet "paste" at all but rather a hard ceramic material.

Calcination The process of heating a raw material to high temperatures, and often red heat, to destroy organic and volatile components which might otherwise promote the evolution of gaseous products that could generate voids or blemishes in a viscous ceramic body paste under the firing process in the kiln. For example, limestone breaks down thermally to lime, calcium oxide, between 650–750 °C and the gaseous carbon dioxide evolved can be trapped in the matrix as unsightly bubbles. Powdered bones were generally heated to destroy the keratotic organic

H. G. M. Edwards, C. J. Brooke, *Coade Stone*, Cultural Heritage Science,
https://doi.org/10.1007/978-3-031-71432-0

components in the hydroxyapatite matrix and coordinated water: the carbon formed initially during the heating process was converted to carbon dioxide at higher temperatures to leave a pure white material. At lower temperatures, artists favoured the heating of ivory or bone to lower temperatures but stopped the process at the formation of carbon, to harvest the "ivory black" or "bone black" pigment which was much appreciated for its depth of colour.

Celadons Celadon is a term that has been used to describe the greenish glazed wares from the Chinese Song Dynasty and later which may be porcelain, stoneware or earthenware that had been fired in a reducing kiln at a high temperature of between 1200 and 1260 °C. Much prized for their beauty, especially in simulation of jade, the distinctive colour of the glaze is ascribed to the chemical reduction of iron (III) oxide (Fe_2O_3) to FeO and its entrapment in the silicaceous matrix. A famous site for celadon production in China was at Longquan. Small amounts of titania confer an attractive yellowish-green tint to the glaze. Celadons were also manufactured in Korea and Thailand. Their popularity waned with the appearance of the fine blue and white decorated porcelains of the Yuan Dynasty.

Ceramic stone From the sixteenth century potters experimented with a type of stoneware that used a white clay, usually a type of ball clay, as a component and fired their pottery at high temperatures rivalling that of soft paste porcelains often up to 1150–1200 °C. The result was a very robust earthenware pottery which after glazing was extremely advantageous for everyday usage: a classic example is *Cologne ware*. In the eighteenth century an adaptation gave rise to *Coade stone* and other formulations which saw multiple usage in architectural projects and decorative large statuary—the re-located *Lion of the South Bank* on Westminster Bridge, London, for instance, was made of Coade stone originally for the Lambeth Brewery in London in 1837 and weighed 14 tons. The defining parameter, of course, is that these ceramic stone composites have no translucency at all and therefore cannot strictly be described as porcelain but they are occasionally referred to historically as *porcellaneous*.

Chemical descriptors Care should be taken to differentiate between the descriptors of chemical compounds and their component elements. For example, silica is SiO_2 and contains silicon and oxygen: likewise, calcia is calcium oxide CaO (lime), alumina is aluminium oxide Al_2O_3, magnesia is magnesium oxide MgO, potash is potassium oxide K_2O and soda is sodium oxide Na_2O. Other chemicals are described precisely as they occur, such as phosphorus pentoxide P_2O_5—also known as phosphorus (V) oxide or phosphorus pentoxide P_2O_5, relative to phosphorus (III) oxide P_2O_3—and sulfur dioxide SO_2. Iron oxide is troublesome as this can mean both ferrous oxide FeO and ferric oxide Fe_2O_3, where the iron is in the oxidation state (II) and (III), respectively, and the former is a green colour and the latter orange-red.

China A generic term for high-fired ceramics, which may be glazed or unglazed, and used for decorative or utilitarian purposes around the home. Believed to originate from descriptors of ceramics that were made in simulation of imported Chinese porcelain, this term covers a wide range of ceramics including porcelain, earthenwares, creamwares, stonewares, celadons, majolica and delftwares. The term "china" has been used freely to describe ceramic articles of a range of compositions since the importation of porcelain from China in the sixteenth century and some of the artefacts do not contain china clay (kaolinite).

Coat-of-arms An officially registered heraldic device granted in Great Britain by Royal authority via the College of Arms in London and the Lord Lyon King of Arms in Scotland to members of a noble family or an arms-bearing organisation. Other European countries have their equivalent officers for the administering and granting of arms. This includes a shield (escutcheon) with dexter (right) and sinister (left) supporters, as viewed from the bearer of the shield—or by an observer from the front of the shield in a mirror image. The supporters are usually real or mythical animals or birds such as the lion, leopard, eagle, unicorn, dragon, wyvern and gryphon. Under the shield (base) a motto may be carried and also a badge of honour. On top of the shield (chief) can sit a coronet signifying the rank in the peerage of the bearer of the arms, a helm and visor accompanied by mantlings (lambrequins) and supporting a twisted silk torse upon which the crest sits. Atop the crest may appear a battle cry legend.

Coefficient of expansion and contraction A parameter which is based upon the ratio of linear change in size of a ceramic material upon the increase or decrease of temperature, usually expressed as a unit of length per degree. It is critically important for ceramics as it results in shrinkage of the initial article upon firing and also for its subsequent use as part of a frieze where the integrity of the structure could be compromised.

Cullet Also known as glass frit, this is a glass additive in ceramic body recipes. Generally purchased by porcelain manufactory proprietors from neighbouring glassworks (for example, as the "end of day glass" mixture) this was then added as a finely ground material to firing process of ceramic syntheses. Cullet is a rather indefinite term scientifically as it could contain either flint glass (a highly refractive glass containing up to 60% lead oxide, PbO) or crown glass (also known as soda glass, which contained either soda or potash, or both, as an alkaline flux, Na_2O and K_2O, but critically for analytical chemical interpretation of the elemental data, this soda glass is lead-free. Analytically the presence of a glass additive is important as in the absence of a lead—containing material from other sources in the recipes, the presence of a lead oxide signature in the analytical data derived from the body paste is a definitive indicator of the presence of a flint glass cullet additive. In several production notes from manufactories, proprietors did not refer to the addition of cullet in their raw materials composition, yet analytically this has been inferred from the presence of a lead oxide signature (from flint glass) or the presence of an abnormally high soda and potash content (arising potentially from a crown, or soda glass cullet additive). Also known as glass frit, a confusion occasionally occurs in terminology because of the use of the word "frit" which is used to also describe a ground calcined body paste after a first firing sequence, whether this contains glass cullet or not.

Elemental oxides Usually these are determined from an SEM/EDAXS or XRF experiment which detects the key elements in fired porcelain such as silicon and aluminium and a range of others: a conversion factor hence needs to be applied to estimate the percentage of silica, alumina or other metal and nonmetal oxide present in the sample. From the ratio of the elemental relative atomic masses in the formula to the corresponding molecular weights of the oxides, the conversion factors can be listed as follows: Si to SiO_2 2.1, Al to Al_2O_3 1.9, Mg to MgO 1.7, Ca to CaO 1.4, Pb to PbO 1.1, Na to Na_2O 1.35, K to K_2O 1.2, Ti to TiO_2 1.7, Fe to Fe_2O_3 1.4 and P to P_2O_5 2.3. This means that an analytical determination of 10% elemental silicon in the experiment would correlate with a content of 21% silicon dioxide (silica) in the specimen. This does not necessarily imply that the ceramic in this case has a composition which required 21% of sand as the analytical result is a compound quantitation of all the components present which contained silica: these might include materials such as sand, flint, chert, smalt, feldspar, glass frit, china clay, secondary ball clays and steatite.

Factory marks impressed, stencilled or cursive script marks which designate the factory or works of origin in a ceramic artefact. These changed with time and ownership and were not universally used on each piece or artefact. In particular, characteristic idiosyncratic cursive script marks can often be attributed or assigned to particular modellers and these can be a fertile area for ceramic historians.

Fake a piece of artwork that has been made to deceive the purchaser into believing that it is something that it is not. Often accompanied by incorrect factory marks, the shapes, sizes, texture of the ceramic body and any decoration are usually also incorrect.

Frit The result of a fine grinding process which is applied to the components of a ceramic paste and especially used to describe the compositional mixtures of bone ash, china clay, flint glass (cullet) and soaprock. The necessity of very fine grinding to produce a homogeneous mixture of components which originally may have very tangible differences in hardness was appreciated by the earliest ceramics manufacturers and a fine frit usually then required just the addition of a single component or water, perhaps with some additional alkaline flux or powdered china clay, to effect a suitable and tractable paste mixture for moulding and firing in a biscuit kiln. The term frit is also applied to a first-stage preparatory firing process of a component mixture which has been fired at a high temperature then ground, mixed with new components and then re-fired to produce biscuit porcelain in the two-stage process that is typical of most porcelain syntheses. Care must be taken in the understanding of the meaning of frit in recipes as it was used to describe both the finely ground body after its initial firing and also the ground glass component used as an additive synthetic raw material prior to its making up as a body paste.

Fusibility The process of melting a ceramic material at elevated temperatures to produce a molten phase with no residual solid residues is affected by the presence of alkaline components which are called fluxes, typically soda, lime, potash and borax. The achievement and stabilisation of this molten phase at high temperatures is essential for the structural integrity of the ceramic body as the appearance of stress cracks can be initiated at lower temperatures at interfacial domains between fusible and infusible materials.

Glass A supercooled vitreous material which is rigid but has not crystallised—glass is often referred to scientifically as a supercooled liquid material. Several raw material components are shared with porcelains, including sand and an alkaline flux comprising calcia, soda and potash. Glass may be crown glass, which is lead-free, or flint glass, which contains significant amounts of lead oxide, which usually has been added to the molten glass in the form of red lead, trilead tetroxide, Pb_3O_4. Pulverised glass residues in the form of cullet, obtained from neighbouring glassworks, who wished to dispose of surplus and residual "end-of-day glass", were frequently added to ceramic raw materials and the determination of an analytical signature for lead oxide in a fired ceramic body results in the conclusion that flint glass residues were a component in the body paste; a crown glass cullet additive is usually manifest in an increased soda, calcia and perhaps also a potash percentage. Glass firing was conducted over a similar temperature range as soft paste porcelains and stonewares especially for their glazing procedure which was broadly accomplished at a temperature of between 950–1150 °C, and the aim was to achieve the cotectic region where the pressure, temperature and composition of the raw material components would result in a melting of the component raw materials without the separate crystallisation out of the individual species. Although normally one would classify a glass as transparent, certain types of glass, such as milk glass, had heavy metal oxides added as opacifiers which rendered them translucent, in keeping with porcelains.

Grog The practice of grinding up damaged and waste fired ceramic items (grog) discarded from the kiln firing of biscuit porcelains and stonewares and re-incorporating them in the paste of new porcelain or stoneware syntheses was adopted in the eighteenth century and frequently involved the purchase by a manufactory of broken porcelains and stonewares from other factories often by the ton weight. Although at first sight this might seem a reasonable procedure to adopt commercially in the use and re-cycling of an otherwise waste material product, it does create a potential difficulty in the modern analytical interpretation of such "mixed" pastes where the presence of unusual and exogenous interlopers in the form of shards and fragments from other factories found archaeologically on site can be potentially registered in what would otherwise constitute an analysis of a standard paste for a particular factory. In this way lead, phosphorus and magnesium could be introduced into a paste analysis for a factory that did not normally utilise these elemental components in raw materials for their recipes—for example if a porcelain grog containing a lead-rich glass was encompassed in the ground frit for a factory that did not normally use a glass in its composite formulation the analytical recording of a signal for lead could be interpreted incorrectly as an experimental change in the compositional recipe for that manufactory.

Holistic approach The consideration of all the factors which might operate for the proper characterisation of materials for provenancing or identity purposes. In a forensic theme this would include historical documentation, witness statements, invoices to clients and personal statements along with scientific analytical data and their interpretation.

Instrumental Chemical Analysis the use of instrumentation to detect qualitatively and to estimate quantitatively the presence of chemical elements and their compounds in ceramic bodies and glazes. The basis of instrumental analysis is the use of special signatures or properties of chemical species or chemical bonds such as silica, SiO_2, or the $Si = O$ bonds in complex macromolecular structures, such as silicaceous ceramic bodies, from their behaviour towards the scattering or absorption of electromagnetic radiation in the electromagnetic spectrum from the low wavelength (high wavenumber) Xray region to the high wavelength (low wavenumber) microwave region. When coupled with a microscope, an analytical instrument is capable of interrogating very small specimens or domain regions in a larger specimen of the order of cubic microns or cubic nanometres. Instrumentation that has been commonly used in the assessment of the constitution of ceramic bodies, glazes and pigments includes scanning electron microscopy/energy dispersive and X Ray spectrometry (SEM/EDAXS), X Ray fluorescence spectroscopy (XRF) for the elemental oxide composition and X Ray diffraction spectrometry (XRD), laser Raman Spectroscopy (RS) and Fourier-transform infrared spectroscopy (FTIR) for mineralogical and pigment identification.

Liquidus The liquidus range refers to the temperature region wherein specific compositions of key elemental oxide components of a ceramic artefact retain their liquid state at elevated temperatures in a firing sequence. A very important concept, it is usually depicted as a region drawn on a ternary diagram with three major components such as silica, alumina and lime /potash /soda expressed as 0–100% (generally in mole fractions). The liquidus region is critical for the stability of a fired ceramic body as a variation in kiln temperature outside of the designated liquidus extremes will invariably result in imperfections arising from crystallisation of the high-temperature moieties produced or a "sagging" of the prototype porcelain piece will occur—either of these affecting the integrity of the resultant artefact and rendering it unfit for purpose. The ternary eutectic point is the minimum temperature at which all three oxide components will retain a liquid phase and from this the preferred composition can be defined.

Minor Additives Ceramics recipes or formulations frequently listed small percentages or amounts of additives such as cobalt blue, or smalt, which was added to remove vestigial traces of yellow colouration in the incipient china clays occurring from small amounts of iron oxides, or ochres. Small amounts of borax, sodium tetraborate decahydrate, to increase fusibility and the plasticity of the body at higher temperatures and arsenic oxide to assist in the generation of a uniform silicaceous phase were common additives to porcelain pastes.

Pigments In ceramics decoration, these are coloured minerals which are thermally stable at the operating temperatures adopted in the glost kiln after application of the glaze slip and firing up to 600 °C or slightly greater. Mineral pigments are usually metal oxides, sulfates and sulfides such as haematite, gypsum and orpiment. Care needs to be taken in the interpretation of old recipes for the decoration of ceramics as minerals were often confusingly assigned an incorrect nomenclature, such as *minium*, which has been used historically to describe both red lead, trilead tetroxide, Pb_3O_4, and cinnabar, mercury sulfide, HgS, and red oxide of lead can mean both red lead ($2PbO.PbO_2$) and litharge (PbO). Analytically, the interrogation of pigments on enamelled porcelains is a fertile area of study as manufacturers often changed their pigment sources and the composition of their pigments with time and the presence of trace quantities of unusual elements such as zinc, niobium, rubidium and copper can indicate valuable information about the timelines of production and sources of the pigment by the manufactory concerned. For example, the presence of a minor elemental component of bismuth in a cobalt blue pigment on an Iznik glazed tile from the sixteenth/seventeenth century indicated that the source of the cobalt ore was European (from the Erzengebirge mine in Saxony) rather than the more usually encountered Asian mines. Arsenic traces in the cobalt blue pigment on Chinese eighteenth century porcelains are indicative of the use of European cobalt ores in the preparation of the pigment rather than the Asian sources which are arsenic—free but which themselves contain manganese.

Porcelain A generic term which historically describes the characteristic hard paste porcelain comprising *kaolinite* and *petuntse* as major raw materials fired at a high kiln temperature; termed "true porcelain" by some historians to differentiate it from "artificial porcelain" which is retained for the soft paste body variants which included synthetic components as raw materials, for example calcined bone ash, potash, lime and glass frit. During the early eighteenth century, the Chinese perfected a soft paste body variant which has been recognised as a comparative rarity. Early Chinese porcelains have a range of compositions which varied both with their production sites geographically and also with the specified quality: the highest quality porcelains from the Imperial kilns at Jingdezhen were composed of an equal ratio of china clay (kaolin) and petuntse.

Raw Materials The raw materials are components comprising the body and glaze recipes used in ceramic syntheses. Mostly these were natural minerals and rocks which were sourced from precise mines and locations which gave rise to several alternative names. Natural raw materials include feldspars, china clays, ball clays and soapstone. Synthetic raw materials include bone ash, pearl ash, soda ash, smalt, lime and magnesia.

Rocks Whereas minerals have a precise and assigned chemical formula and established crystal structures with recognised polymorphs (different three-dimensional spatial structures for the same chemical formulation, e.g. calcite and aragonite, both of which are formulated chemically as $CaCO_3$), rocks are mixtures or aggregates of minerals with ill-defined chemical compositions, such as soapstone, which is often represented in its purest form as steatite, a hydrated magnesium silicate in its mineral form of talc, $Mg_3Si_4O_{10}(OH)_2$. Care must be taken, however, since even minerals may be found impure or containing associated species: a good example is *lapis lazuli*, the semi-precious gemstone much used as a blue pigment in art works which can be found associated with minor components of pyrites, FeS_2, or calcite, $CaCO_3$, depending upon its geological and geographical source—generally *lapis lazuli* from Afghanistan has inclusions of pyrites whereas that sourced from the Chilean mines contains calcite. Forensically, the presence of trace elements can be a good indicator of the source of a particular raw material used in ceramic syntheses.

Saggar An unglazed ceramic stand or feet for separating moulded pieces stacked together in a kiln prior to firing.

Shard Or "potsherd", is a broken piece of porcelain or pottery, usually found in waste dumps near ceramic kilns and formed from imperfect, commercially useless pieces which have been destroyed after firing. Generally unglazed, they are occasionally found glazed and decorated, for example, in archaeological excavations of ceramic factories or living quarters.

Slaked lime The product of the calcination of limestone (calcium carbonate) or dolostone (dolomite or dolomitised limestone) which results in the gaseous evolution of carbon dioxide and residual lime (calcium oxide) and/or magnesia. Reaction of this product with water gives calcium and/or magnesium hydroxide, the so-called slaked lime or limewash, used as a white base coat for buildings or limewash putty used as a filler and mortar. On reacting with moist air, the lime or magnesia gives the respective carbonate, namely calcite or dolomite, which forms as a hard surface skin so preventing further substrate reaction.

Trace elements The detection of minor components of elemental oxides in fired ceramics is an important aspect of modern analytical chemistry as it affords a unique route into the determination of the sources of the raw materials used in the recipes and formulations. Trace element detection in the range of ppm (parts per million, where 1 ppm is equivalent to 0.0001%) has been used to define changes made in manufactory sources of key raw materials such as clays and to enlarge upon the chronological use of pigments for porcelain decoration, thereby exposing modern attempts to fake antique ceramics by the use of incompatible pigments. Examples of these trace elements in ceramics which have assisted in this respect are zirconium, strontium, bismuth, rubidium, titanium, boron, cadmium, niobium, copper and zinc.

Translucency This is perhaps the greatest achievable asset of porcelain manufacture to which every manufacturer subscribed and hoped to attain in emulation of the Chinese "eggshell" wares which were imported to Europe in the mid-eighteenth century. It describes the transmission of visible radiation, or light, through a solid object: at one end of the spectral scale is glass, which is usually *transparent* (100% transmission) and at the other is earthenware which is *opaque* (0% transmission). However, see entry above for "Glass", where the occurrence of a milk glass is noted in which the transparency is less than 100% and the glass is then termed translucent. Porcelain is measured by its translucency, which is the clarity for transmission when viewed with background lighting. An intermediate descriptor is "semi-opaque" or even "semi-porcelain", which was applied to some china and stonewares during the mid-nineteenth century, which is rather indefinite and conveys little information about their true category. Other descriptors such as "vitreous china" and "ironstone china" likewise do little to assist in their classification. By definition, porcelain must be able to transmit light to some extent unlike stonewares, faience, terracotta and earthenwares.

Index

H. G. M. Edwards, C. J. Brooke, *Coade Stone*, Cultural Heritage Science,
https://doi.org/10.1007/978-3-031-71432-0

Milton Keynes UK
Ingram Content Group UK Ltd.
UKHW020712021124
450420UK00001B/12